"十四五"职业教育国家规划教材

"十三五"江苏省高等学校重点教材（2018-2-176）

物联网移动应用开发

主　编　季云峰　刘　丽

副主编　冯立元

参　编　匡　亮　高　云

机械工业出版社

本书以物联网智慧工厂移动端应用系统开发为基础,将开发过程拆解成 15 个任务,任务安排既遵循 Android 知识点的学习路径,又尽量符合移动应用开发的逻辑过程,每个任务完成后都可以编译、运行,后续任务在前一任务的基础上进行扩展,最终完成整个应用的开发。读者通过重构、复原该系统来掌握物联网移动应用开发的常用技术,熟悉开发的过程。

本书适合作为高职院校电子信息类、计算机类相关专业的教材,也适合对物联网移动应用开发感兴趣的读者阅读学习。

本书配有授课电子课件,需要的教师可登录 www.cmpedu.com 免费注册,审核通过后下载,或联系编辑索取(QQ:13261377872,电话:010-88379739)。

图书在版编目(CIP)数据

物联网移动应用开发 / 季云峰,刘丽主编. —北京:机械工业出版社,2020.1(2025.1 重印)
"十三五"江苏省高等学校重点教材
ISBN 978-7-111-64720-1

Ⅰ. ①物… Ⅱ. ①季… ②刘… Ⅲ. ①移动终端-应用程序-程序设计-高等学校-教材 Ⅳ. ①TN929.53

中国版本图书馆 CIP 数据核字(2020)第 024098 号

机械工业出版社(北京市百万庄大街 22 号 邮政编码 100037)
策划编辑:王海霞　　责任编辑:王海霞　车　忱
责任校对:张艳霞　　责任印制:单爱军

北京虎彩文化传播有限公司印刷

2025 年 1 月·第 1 版·第 11 次印刷
184mm×260mm·15 印张·370 千字
标准书号:ISBN 978-7-111-64720-1
定价:49.90 元

电话服务　　　　　　　　　　　网络服务
客服电话:010-88361066　　　　机　工　官　网:www.cmpbook.com
　　　　　010-88379833　　　　机　工　官　博:weibo.com/cmp1952
　　　　　010-68326294　　　　金　书　网:www.golden-book.com
封底无防伪标均为盗版　　　　机工教育服务网:www.cmpedu.com

高等职业教育系列教材
计算机专业编委会成员名单

名誉主任 周智文

主　　任 眭碧霞

副 主 任 林　东　王协瑞　张福强　陶书中　龚小勇
　　　　　　王　泰　李宏达　赵佩华　刘瑞新

委　　员（按姓氏笔画排序）
　　　　　　万　钢　万雅静　卫振林　马　伟　王亚盛
　　　　　　尹敬齐　史宝会　宁　蒙　朱宪花　乔芃喆
　　　　　　刘本军　刘贤锋　刘剑昀　齐　虹　江　南
　　　　　　安　进　孙修东　李　萍　李　强　李华忠
　　　　　　李观金　杨　云　肖　佳　何万里　余永佳
　　　　　　张　欣　张洪斌　陈志峰　范美英　林龙健
　　　　　　林道贵　郎登何　胡国胜　赵国玲　赵增敏
　　　　　　贺　平　袁永美　顾正刚　顾晓燕　徐义晗
　　　　　　徐立新　唐乾林　黄能耿　黄崇本　傅亚莉
　　　　　　裴有柱

秘 书 长 胡毓坚

关于"十四五"职业教育
国家规划教材的出版说明

为贯彻落实《中共中央关于认真学习宣传贯彻党的二十大精神的决定》《习近平新时代中国特色社会主义思想进课程教材指南》《职业院校教材管理办法》等文件精神，机械工业出版社与教材编写团队一道，认真执行思政内容进教材、进课堂、进头脑要求，尊重教育规律，遵循学科特点，对教材内容进行了更新，着力落实以下要求：

1. 提升教材铸魂育人功能，培育、践行社会主义核心价值观，教育引导学生树立共产主义远大理想和中国特色社会主义共同理想，坚定"四个自信"，厚植爱国主义情怀，把爱国情、强国志、报国行自觉融入建设社会主义现代化强国、实现中华民族伟大复兴的奋斗之中。同时，弘扬中华优秀传统文化，深入开展宪法法治教育。

2. 注重科学思维方法训练和科学伦理教育，培养学生探索未知、追求真理、勇攀科学高峰的责任感和使命感；强化学生工程伦理教育，培养学生精益求精的大国工匠精神，激发学生科技报国的家国情怀和使命担当。加快构建中国特色哲学社会科学学科体系、学术体系、话语体系。帮助学生了解相关专业和行业领域的国家战略、法律法规和相关政策，引导学生深入社会实践、关注现实问题，培育学生经世济民、诚信服务、德法兼修的职业素养。

3. 教育引导学生深刻理解并自觉实践各行业的职业精神、职业规范，增强职业责任感，培养遵纪守法、爱岗敬业、无私奉献、诚实守信、公道办事、开拓创新的职业品格和行为习惯。

在此基础上，及时更新教材知识内容，体现产业发展的新技术、新工艺、新规范、新标准。加强教材数字化建设，丰富配套资源，形成可听、可视、可练、可互动的融媒体教材。

教材建设需要各方的共同努力，也欢迎相关教材使用院校的师生及时反馈意见和建议，我们将认真组织力量进行研究，在后续重印及再版时吸纳改进，不断推动高质量教材出版。

机械工业出版社

前　言

　　党的二十大报告指出，加快建设国家战略人才力量，努力培养造就更多大师、战略科学家、一流科技领军人才和创新团队、青年科技人才、卓越工程师、大国工匠、高技能人才。为适应当前高职院校师生教学项目化和任务化的需求，本书尝试用一个经过改造的物联网智慧工厂移动端应用系统来覆盖整个移动应用开发课程的教学过程，通过将该系统分解成 15 个任务来支持项目化、模块化教学的需求。每个任务都设定了知识目标和技能目标，实现对 Android 应用开发知识点和技能点的覆盖。系统使用 Android Studio 来开发，读者完成每个任务后都可以编译、部署、运行应用，可以直观地看到学习的成果，后续任务在前一任务的基础上进行扩展，最终完成整个系统的开发。

　　本书是一本介绍物联网方向 Android 移动应用开发的教材，不是一本系统介绍 Android 应用开发的教材，因此在教材中部分常用的 Android 开发知识点没有介绍，如 Fragment、Service、ContentProvider 等，读者可以根据自己的需求进行拓展学习。

　　本书是"移动应用开发"在线开放课程的配套教材，读者可以通过中国大学 MOOC 网站加入在线开放课程的学习。

　　本书是"十三五"江苏省高等学校重点教材，由江苏信息职业技术学院季云峰、刘丽、冯立元、匡亮、高云编写，季云峰、刘丽任主编，冯立元任副主编，匡亮、高云参编。

　　在教材编写的过程中，得到了众多同行老师的关心和指导，得到了北京新大陆时代教育科技有限公司和高等职业教育系列教材计算机专业编委会的鼎力帮助和支持。

　　由于编者水平有限，时间仓促，尽管我们尽了最大的努力，但书中仍难免有不妥和错误之处，恳请读者批评指正。

编　者

目 录

前言
任务 1 系统概述及设计 ……………… 1
1.1 项目背景 ……………………………… 1
1.2 项目方案 ……………………………… 1
1.3 系统部署 ……………………………… 3
1.4 系统功能 ……………………………… 3
1.4.1 系统登录和注册 …………………… 3
1.4.2 系统主界面 ……………………… 4
1.4.3 全局参数设置 …………………… 4
1.4.4 传感器历史数据显示 ……………… 4
1.4.5 禁入区域警报数据 ……………… 6
1.4.6 禁入区域摄像监控 ……………… 6
1.4.7 抽屉导航菜单 …………………… 6
任务 2 创建开发环境和项目 ………… 9
2.1 初识 Android ………………………… 9
2.2 Android 平台架构 ………………… 10
2.3 创建开发环境和工程 ……………… 11
2.3.1 创建开发环境 …………………… 11
2.3.2 创建工程 ………………………… 12
2.4 运行应用 …………………………… 23
2.5 更改应用的启动图标和应用名称 …………………………………… 30
任务 3 创建 Splash 界面 …………… 34
3.1 创建 Splash 活动和布局 ………… 34
3.2 编辑 Splash 布局 ………………… 36
3.3 编辑 Splash 活动 ………………… 39
3.4 修改活动及其生命周期 …………… 45
任务 4 创建系统主界面 ……………… 53
4.1 选择主界面布局方式 ……………… 54
4.2 创建线性布局 ……………………… 54
4.2.1 添加环境监控布局 ……………… 54
4.2.2 添加禁入区域监控布局 ………… 57
4.2.3 添加设备控制布局 ……………… 58
任务 5 使用活动条导航到全局参数设置界面 ……………………………… 62
5.1 添加活动条和主题 ………………… 62
5.2 创建动作项 ………………………… 64
5.2.1 在菜单资源文件中定义动作项 …… 64
5.2.2 在活动中实现 onCreateOptionsMenu() 方法 ……………………………… 65
5.2.3 用 onOptionsItemSelected() 方法响应活动条单击 ……………………… 66
任务 6 创建全局参数设置界面 ……… 70
6.1 添加网格布局 ……………………… 70
6.2 添加按钮单击事件 ………………… 82
6.2.1 通过匿名内部类实现 …………… 82
6.2.2 通过独立类实现 ………………… 82
6.2.3 通过 OnClickListener 接口实现 …… 83
6.3 保存全局参数 ……………………… 83
6.3.1 使用 SharedPreference 保存参数 …… 84
6.3.2 使用用户自定义 Application 保存全局参数 ……………………… 90
任务 7 从云平台获取传感器数据并显示 …………………………………… 100
7.1 使用第三方提供的 jar 包 ………… 101
7.2 添加网络权限 ……………………… 103
7.3 创建 CloudHelper 帮助类 ………… 105
7.4 从云平台获取传感器数据并在主界面更新 ……………………………… 107
7.4.1 通过 Handler 机制实现线程消息传递 ……………………………… 108
7.4.2 使用定时器定时更新主界面数据 …… 109
任务 8 通过云平台控制执行器 ……… 114
8.1 创建执行器控制方法 ……………… 114
8.2 使用适配器设置执行器控制状态 ………………………………… 115
8.3 使用 setResult 和 onActivityResult 机制实现返回 ……………………… 120
任务 9 创建执行器状态动画 ………… 123

9.1 创建通风控制系统风扇动画 ……… 123
9.2 创建空调控制系统送风动画 ……… 127
9.3 创建照明控制系统灯光动画 ……… 129

任务 10 绘制传感器数据折线图 ……… 135
10.1 使用 SQLite 数据库保存数据 …… 136
10.2 创建 SQLite 帮助器 ……………… 137
10.3 使用 MPAndroidChart 来绘制
　　　传感器数据折线图 ……………… 141
　　10.3.1 导入 MPAndroidChart 图表库 …… 141
　　10.3.2 创建活动 DataChartActivity ……… 142

**任务 11 存储报警信息至服务器并创建
　　　　警报数据界面** ……………… 147
11.1 创建和部署 WebService ………… 148
　　11.1.1 创建 WebService ……………… 149
　　11.1.2 部署 WebService ……………… 151
11.2 创建 WebServiceHelper 类 ……… 152
11.3 更新活动 MainActivity 中的
　　　定时器任务 ……………………… 154
11.4 查看历史报警信息 ……………… 155
　　11.4.1 为 ListView 创建布局 ………… 156
　　11.4.2 创建自定义适配器 WarnAdapter …… 158
　　11.4.3 创建活动 WarnListActivity ……… 161

任务 12 创建摄像头监控界面 ……… 163
12.1 创建摄像头监控布局文件 ……… 163
　　12.1.1 创建摄像头布局文件 …………… 164
　　12.1.2 创建摄像头控制按钮布局文件 …… 166
12.2 创建 HTTP 访问类
　　　HttpRequest ……………………… 170

12.3 实现摄像头访问 ………………… 173

任务 13 创建抽屉导航 ……………… 179
13.1 使用 ToolBar、DrawLayout 和
　　　NavigationView 创建抽屉导航 … 179
　　13.1.1 使用 ToolBar 组件 ……………… 181
　　13.1.2 创建导航栏 ……………………… 181
13.2 创建抽屉导航界面中各功能
　　　模块 ……………………………… 185

任务 14 创建登录和注册功能 ……… 203
14.1 创建并部署 WebService ………… 204
14.2 在 WebServiceHelper 类中添加
　　　登录和注册功能 ………………… 206
14.3 创建登录和注册界面 …………… 209
　　14.3.1 创建登录界面 …………………… 209
　　14.3.2 创建注册界面 …………………… 213
14.4 创建 LoginActivity 活动实现登录
　　　功能 ……………………………… 217
14.5 创建 RegisterActivity 活动实现
　　　注册功能 ………………………… 219

任务 15 实现多语言切换 …………… 222
15.1 更新语言选择界面 ……………… 222
　　15.1.1 创建语言选择界面 ……………… 223
　　15.1.2 创建 ListView 适配器 …………… 224
15.2 简体中文和英文语言适配 ……… 226
15.3 实现 Android 应用内切换
　　　语言 ……………………………… 227

参考文献 …………………………… 232

任务 1 系统概述及设计

任务概述
本任务主要介绍物联网智慧工厂移动端应用系统的设计。
知识目标
- 了解物联网系统设计。
- 了解物联网系统的部署。
- 了解物联网云平台。

技能目标
- 能绘制系统拓扑图。
- 能部署物联网系统。

1 系统概述及设计

1.1 项目背景

物联网智慧工厂监控系统移动端应用开发是某公司工厂监控管理改造的 Android 移动端项目。该系统可以利用智能移动终端实时感知工厂内部温度、湿度、光照度等环境信息，通过预设的阈值，使通风系统、照明系统等联动，实现工厂内部温度、湿度、光照度的自动调节；系统可以实时感知各个关键非工作区域（危险区域）有人闯入的情况，如果有人闯入，系统自动推送报警信息到移动端；可以通过视频监控系统远程实时查看工厂现场生产作业情况。

1.2 项目方案

物联网智慧工厂监控管理系统硬件拓扑结构如图 1-1 所示。

1）光照传感器、温湿度传感器等接入 Zigbee 节点模块，实现光照度、温湿度模拟数据的采集。

2）人体红外传感器、排气扇继电器等接入 ADAM-4150 数字量采集器，实现人体红外信息的采集和对排气扇的开关控制。

3）照明灯接入带有继电器的 Zigbee 模块，实现对照明灯的开关控制。

4）ADAM-4150 通过 485 总线接入到物联网数据采集网关，各 Zigbee 节点通过 Zigbee 网络接入到物联网数据采集网关。

5）物联网数据采集网关通过 WiFi 网络接入 Internet，连接到部署在公网的物联网云平台。

6）移动端（智能手机、平板电脑等）通过物联网云平台实现远程监控和管理。

项目开发中以排气扇 1 替代模拟通风系统执行器，排气扇 2 模拟空调系统执行器，照明灯 1 模拟照明系统执行器。

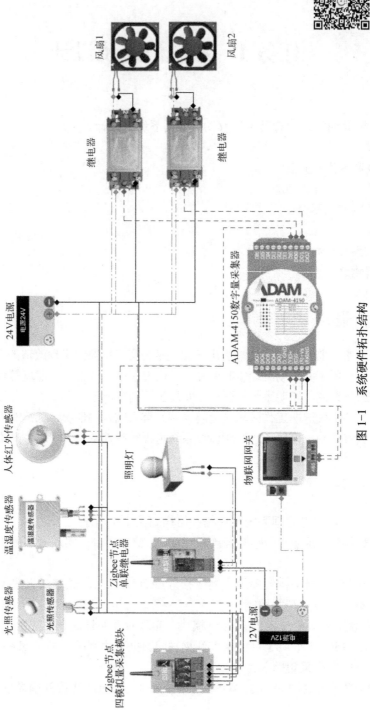

图 1-1 系统硬件拓扑结构

1.3 系统部署

读者可以基于新大陆实训设备（NLE-JS2000）和私有云进行软硬件系统的安装和部署，也可以使用新大陆虚拟仿真软件，利用新大陆公司免费开放的公网物联网云平台（http://www.nlecloud.com）完成系统的搭建。

2 使用虚拟仿真软件进行系统搭建

系统安装部署步骤如下。
1）Zigbee 节点的烧写和组网，各硬件设备的连接。
2）物联网云平台（私有云）的部署、配置。
3）物联网数据采集网关的配置。
4）服务端程序（WebService）的部署。

详细步骤参看本课程的在线开放课程。打开中国大学 MOOC 网站（http://www.icourse163.org），搜索"移动应用开发"，或在浏览器中输入 http://www.icourse163.org/course/JSIT-1001754058，进入该课程的学习。用户需要注册方可进入学习。

（扫描二维码进入学习）

1.4 系统功能

1.4.1 系统登录和注册

系统提供了用户注册和登录的功能，用户信息保存在服务器端，应用通过 WebService 服务从服务器端远程保存和读取用户信息，如图 1-2 和图 1-3 所示。

图 1-2 登录界面

图 1-3 注册界面

1.4.2 系统主界面

系统主界面实时显示工厂环境信息和禁入区域的监控情况，单击实时数据视图可以进入历史数据查看界面。在设备控制区，用户可以手动打开或关闭 3 个系统的执行器，也可将执行器设为自动控制，并以动画显示指示执行器的打开和关闭。在主界面中通过活动条可以导航到全局参数设置和摄像监控界面，如图 1-4 和图 1-5 所示。

图 1-4 系统主界面

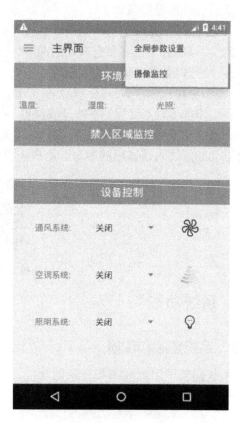

图 1-5 导航菜单

1.4.3 全局参数设置

全局参数设置界面用于设置物联网云平台、监控摄像设备、各类传感器和执行器的相关参数。保存全局参数后，可以从物联网云平台获取传感器的实时数据并显示，这时用户可以设置执行器的控制状态。传感器数据每隔 5s 保存至本地 SQLite 数据库中，如图 1-6 和图 1-7 所示。

1.4.4 传感器历史数据显示

在主界面单击实时数据视图可以查看各传感器的历史数据，以折线图方式显示，如图 1-8 和图 1-9 所示。

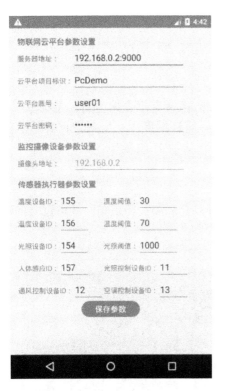

图 1-6 全局参数设置

图 1-7 系统主界面

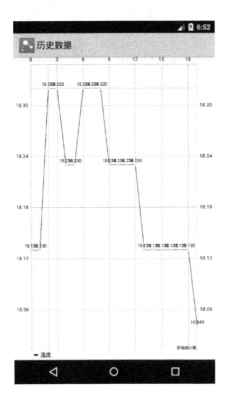

图 1-8 温度数据折线图

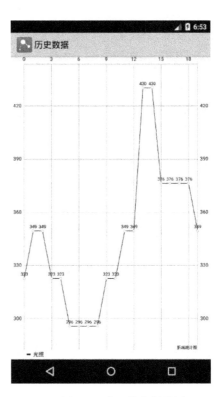

图 1-9 光照数据折线图

1.4.5 禁入区域警报数据

在主界面中单击禁入区域情况视图，可以打开禁入区域警报数据界面。系统每隔 5s 从物联网云平台获取人体传感器数据并保存到远程服务器数据库中，警报数据界面从服务器 SQL Server 数据库中获取"有人闯入"的信息并显示，如图 1-10 所示。

1.4.6 禁入区域摄像监控

通过网络摄像头实现禁入区域摄像监控。用户可以开启和关闭摄像，并通过方向键来调整监控角度，如图 1-11 所示。

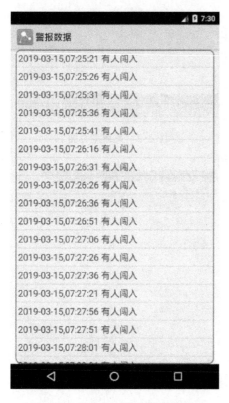

图 1-10　禁入区域警报数据界面

图 1-11　禁入区域摄像监控界面

1.4.7 抽屉导航菜单

抽屉导航菜单包含了个人设置、语言选择（包含简体和繁体中文）、软件信息、切换账户、退出程序等功能，如图 1-12～图 1-17 所示。

图 1-12　抽屉导航菜单

图 1-13　个人设置界面

图 1-14　语言选择界面

图 1-15　软件信息界面

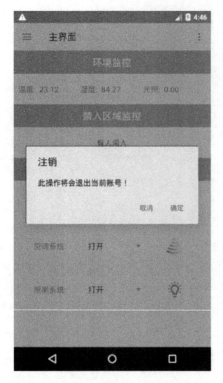 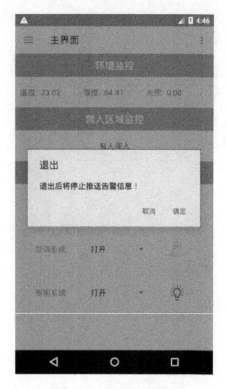

图 1-16　切换账户界面　　　　　图 1-17　退出程序界面

任务 2　创建开发环境和项目

任务概述

本任务要完成 Android Studio 开发环境的创建，在 Android Studio 中完成 SmartFactory 工程项目的创建、配置、运行，并实现应用启动图标和应用名称的修改。

知识目标

- 了解 Android 平台架构。
- 了解 Android API 版本。
- 掌握活动（Activity）和布局（Layout）概念。
- 掌握 Android Studio 中的工程目录。
- 了解 Gradle 系统构建工具。
- 了解 Android 虚拟设备（AVD）。
- 了解 Android 应用运行过程。

技能目标

- 能搭建 Android Studio 开发环境。
- 能完成 Android Studio 工程项目的创建、配置和运行。
- 能完成应用启动图标和名称的修改。

3　初识 Android 平台

2.1　初识 Android

在开始任务之前首先要了解一下 Android 这个平台。Android 由 Google 发起，是一个基于 Linux 的全面的开源平台。它作为一个强大的开发框架，包含了结合 Java 和 XML 构建应用所需的全部特性。更重要的是基于这个平台，可以将应用部署到各类不同的智能设备上。目前使用 Android 的设备已经达到数十亿台。

一个典型的 Android 应用的组成包含布局、活动和资源。

布局（Layout）定义了应用的外观，各个屏幕看上去是什么样。布局通常由 XML 定义，主要包含一些 GUI 组件，如按钮、文本和标签等。

活动（Activity）定义了应用要做些什么，它使用 Java 编写，是一种特殊的 Java 类。

资源（Resource）包含了应用需要的额外的资源，如图片文件、声音文件、数据文件等。

由此可见，Android 应用实际上是由一些目录中的一组文件构成。构建应用时，所有这些文件会"捆绑"在一起，构成一个应用（apk 文件），最终被安装在设备上运行。

2.2 Android 平台架构

Android 平台由很多不同的组件构成，包括一些系统应用（如通信录），一组 API 以及大量支持文件和库，这些 API 可以帮助用户控制应用的外观和行为。这里只对 Android 平台做简单的介绍，后续章节会对 Android 平台包含的组件做详细解释。平台如图 2-1 所示，包括以下几部分。

图 2-1 Android 软件栈

1）系统应用（System Apps）：Android 提供了一组核心应用，如电话、电子邮件、相机、日历等。

2）应用框架（Java API Framework）：用户构建应用时可以使用这些系统应用所使用的框架 API，这些 API 使用 Java 语言编写，包括以下组件和服务。

① 丰富、可扩展的视图系统，可用以构建应用的 UI，包括按钮、列表、文本框，甚至可嵌入的网络浏览器。

② 资源管理器，用于访问非代码资源，如本地化的字符串、图形和布局文件。

③ 通知管理器，可让所有应用在状态栏中显示自定义提醒。

④ Activity 管理器，用于管理应用的生命周期，提供常见的导航返回栈。

⑤ 内容提供者，可让应用访问其他应用（如"联系人"应用）中的数据或者共享其自己的数据。

3）库（Native C/C++ Libraries）：应用框架底层有一组 C 和 C++编写的原生库，可以通过框架 API 访问这些库。如果开发的是需要 C 或 C++ 代码的应用，可以使用 Android NDK 直接从原生代码访问某些原生平台库。

4）Android 运行时环境：Android 运行时环境提供了一组核心库，实现了 Java 编程语言的大部分功能。所有 Android 应用都在自己单独的进程中运行。

5）硬件抽象层（Hardware Abstraction Layer）：对硬件设备的抽象和封装，为 Android 在不同硬件设备上提供统一的访问接口。

6）Linux 内核（Kernel）：最底层的是 Linux 内核，Android 平台的基础是 Linux 内核，Android 依赖这个内核提供驱动以及核心服务（如安全和内存管理）。

2.3 创建开发环境和工程

创建一个基本的 Android 应用需要完成以下三件事。
- 建立一个开发环境。
- 构建一个基本应用。
- 在 Android 模拟器中运行这个应用。

2.3.1 创建开发环境

Java 是开发 Android 应用的最流行的语言。Android 设备并不会运行.class 和.jar 文件。实际上，为了提高速度和电池性能，Android 设备会使用自己的优化格式表示编译代码。这就说明不能使用常规的 Java 开发环境，还需要有特殊的工具把编译的代码转换成 Android 格式，另外要把应用部署到一个 Android 设备上，以便运行后进行调试。为实现这些功能，需要用到 Android SDK。

1．Android SDK

Android 软件开发包（Software Development Kit，SDK）包含开发 Android 应用所需的库和工具。Android SDK 主要是以 Java 语言为基础，用户可以使用 Java 语言来开发 Android 平台上的软件应用。Android SDK 安装目录如图 2-2 所示。

图 2-2 Android SDK 安装目录

安装目录中包含了以下主要部分。

1）build-tools：编译工具目录，包含了转化为 Dalvik 虚拟机的编译工具。

2）extras：某些扩展插件。
3）platforms：存放 Android 不同版本的 API。
4）platform-tools：存放一些通用工具，如 adb.exe 等。
5）sources：SDK 的源代码。
6）system-images：系统镜像（模拟器的镜像文件）。
7）tools：编程时使用的一些重要工具，如 Android 调试工具 DDMS 等。

4　Android Studio 下载与安装

2．Android Studio

IntelliJ IDEA 是完成 Java 开发的最流行的 IDE 之一。Android Studio 是 IDEA 的一个特殊版本，其中包括了一个 Android SDK 以及额外的一些 GUI 工具来帮助实现移动应用开发。

除了提供编辑器以及允许访问 Android SDK 中的工具和库之外，Android Studio 还提供了一些模板，可以用来创建新的应用和类，利用 Android Studio 可以很容易地打包和运行应用。

Android Studio 是一个 Java 开发环境，所以需要确保计算机上安装了正确的 Java 版本。可以到 Oracle 网站下载 JDK，下载网址有时会改变，可以在网上搜索一下。

https://www.oracle.com/technetwork/java/javase/downloads/index.html

安装完 JDK 后，可以到 Google 开发者网站下载 Android Studio。

https://developer.android.google.cn/studio

如果 Google 开发者网站无法访问，可以到 Android Studio 中文社区（www.android-studio.org）下载安装。本书使用 Android Studio 3.2 版本开发。

5　创建工程

2.3.2　创建工程

完成开发环境的创建后，打开 Android Studio 可以看到欢迎界面，在此就可以创建 Android 应用了，如图 2-3 所示。

图 2-3　Android Studio 欢迎界面

1．配置工程

在配置工程界面，需要指定工程项目的名称、使用哪个公司域名、项目的存储目录，以及包名。

Android Studio 使用公司域名和应用名来构成应用使用的包名。例如，应用名（Application name）为"SmartFactory"，公司域名（Company domain）为"jsit.edu.cn"，Android Studio 就会自动生成包名（Package name）cn.edu.jsit.smartfactory。在 Android 中包名非常重要，Android 设备将用这个包名来唯一表示应用，管理同一个应用的多个版本，在整个生命周期中包名要保持不变。

如图 2-4 所示，输入相应信息，单击"Next"（下一步）按钮。

图 2-4　创建 Android 工程项目

2．选择应用设备类型和最低支持版本

Android 应用设备类型包括 5 类，默认选择"Phone and Tablet"。默认最低 SDK 版本为"API 17:Android 4.2(Jelly Bean)"，表示 API 层次为 17，Android 版本为 4.2（果冻豆），如图 2-5 所示。

其中，Phone and Tablet 表示应用是一个手机和平板项目。Wear OS 表示应用是一个可穿戴设备（如手表）项目。TV 表示应用是一个 Android TV 项目。Android Auto 表示应用是一个汽车项目（其需要连接手机使用）。Android Things 表示应用是一个嵌入式设备（如树莓派 3B）项目。

最低 SDK 版本是应用支持的最低版本，应用将在这个版本或者更高版本 API 的设备上运行，如果设备的 API 版本比它低，应用就无法运行。

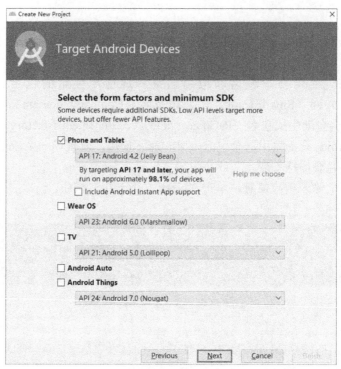

图 2-5 选择应用设备类型和最低 SDK 版本

Android 版本和 API 层次对应关系如表 2-1 所示。

表 2-1 Android 版本和 API 层次对应关系

Android 版本	英文代号	中文代号	API 层次
1.0			1
1.1	Petit Four		2
1.5	CupCake	纸杯蛋糕	3
1.6	Donut	甜甜圈	4
2.0/2.1	Eclair	闪电泡芙	5～7
2.2	Froyo	冻酸奶	8
2.3-2.3.2 2.3.3-2.3.7	Gingerbread	姜饼	9、10
3.0/3.1/3.2	Honeycomb	蜂巢	11、12、13
4.0-4.0.2 4.0.3-4.0.4	Ice Cream Sandwich	冰激凌三明治	14、15
4.1/4.2/4.3	Jelly Bean	果冻豆	16、17、18
4.4	KitKat	奇巧巧克力棒	19
5.0/5.1	Lollipop	棒棒糖	21、22
6.0	Marshmallow	棉花糖	23
7.0/7.1	Nougat	牛轧糖	24、25
8.0/8.1	Oreo	奥利奥	26、27
9.0	Pie	馅饼	28
10.0	Android 10		29

开发 Android 应用要仔细考虑应用与哪些版本兼容。如果指定应用与最新版本兼容，如 8.1 Oreo API 27，通过图 2-6 可以看到这个版本仅能在 1.1%的设备上运行。为了应用能够在绝大多数 Android 设备上运行，这里选择 4.2 Jelly Bean API 17。

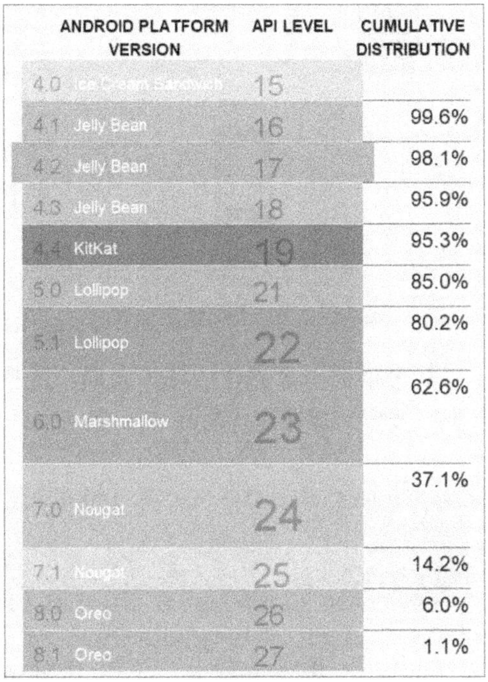

ANDROID PLATFORM VERSION	API LEVEL	CUMULATIVE DISTRIBUTION
4.0 Ice Cream Sandwich	15	
4.1 Jelly Bean	16	99.6%
4.2 Jelly Bean	17	98.1%
4.3 Jelly Bean	18	95.9%
4.4 KitKat	19	95.3%
5.0 Lollipop	21	85.0%
5.1 Lollipop	22	80.2%
6.0 Marshmallow	23	62.6%
7.0 Nougat	24	37.1%
7.1 Nougat	25	14.2%
8.0 Oreo	26	6.0%
8.1 Oreo	27	1.1%

图 2-6　设备版本数量分布

3．创建活动和布局

所有的应用都是由一个个不同的屏幕构成的集合。每一个屏幕都由一个活动和一个布局所构成。

活动（Activity）通常关联一个屏幕，通过与用户来交互完成某项任务，如一个活动要写 E-mail、照相或者验证用户名和密码。活动中所有操作都与用户密切相关，是一个负责与用户交互的组件，用 Java 编写。

布局（Layout）描述了屏幕的外观。布局通常写在一个 XML 文件中，它告诉 Android 屏幕上的按钮、文本框、图像等不同 GUI 组件如何组织。

总之，布局定义了如何表示用户界面，活动定义了如何动作。

下面详细分析活动和布局如何共同创建一个用户界面，如图 2-7 所示。

① 设备启动应用，创建一个活动对象。
② 这个活动对象指定一个布局。
③ 活动告诉 Android 在屏幕上显示这个布局。
④ 在设备上显示包含这个布局的用户界面。
⑤ 活动通过运行应用代码对交互做出响应。
⑥ 活动更新页面。
⑦ 用户将在设备上看到更新后的页面。

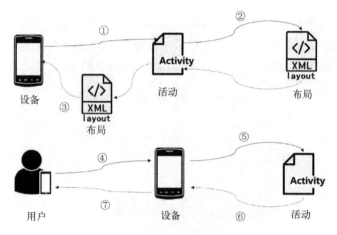

图 2-7 用户界面创建过程

Android Studio 提供了 13 个模板用来创建活动和布局。选择"Empty Activity"（空活动），再单击"Next"（下一步）按钮，如图 2-8 所示。

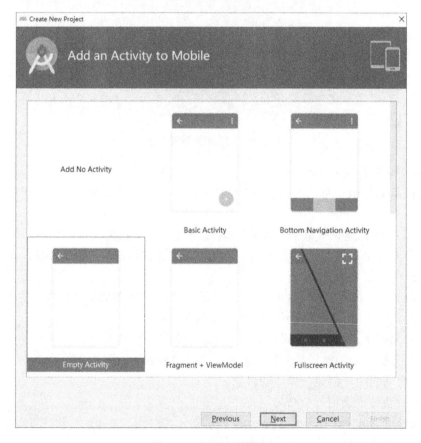

图 2-8 选择活动模板

4．配置活动

输入活动的名称为"MainActivity"，勾选"Generate Layout File"（生成布局文件）复选

框。活动是一个 Java 类，因此将自动创建一个 MainActivity.java 类文件。

输入布局名称"activity_main"，取消勾选"Backwards Compatibility（AppCompat）"复选框，该选项用于添加 AppCompat 包提供向后兼容性，如图 2-9 所示。布局是 XML 文件，因此将自动创建一个 activity_main.xml 的 XML 文件。

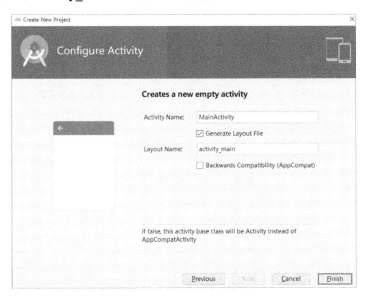

图 2-9　选择活动模板

接下来，Android Stuido 会自动构建应用，构建应用对于初学者来说是一个复杂的概念，这里暂不详述。

单击"Finish"（完成）按钮之后，进入 Android Studio 工程界面，如图 2-10 所示。

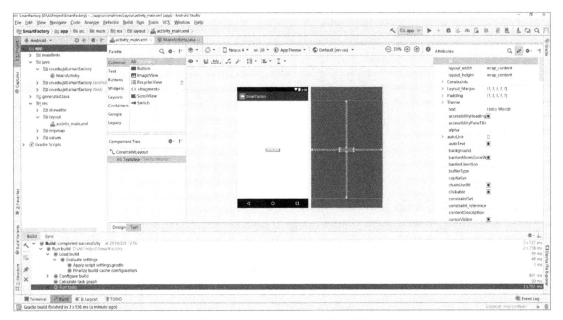

图 2-10　Android Studio 工程界面

17

5. 工程目录

Android 应用实际上就是使用文件夹结构组织的一系列文件的集合。Android Studio 会自动创建这些文件和结构，单击左侧的"Project"（项目）选项，可以浏览整个项目包含的文件夹和各类文件，如图 2-11 所示。

6　工程目录和 Gradle

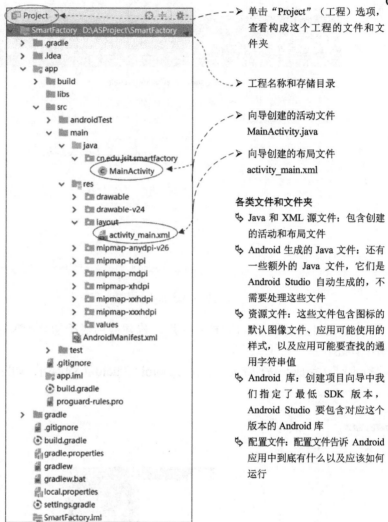

图 2-11　工程目录

6. Gradle

Android Studio 工程使用 Gradle 构建系统来编译和部署应用。Gradle 工程有一个标准布局。Gradle 是一个基于 Apache Ant 和 Apache Maven 概念的项目自动化构建工具，用于管理应用的依赖、打包、部署、发布以及差异管理等。

Project 下面的 build.gradle 文件是整个项目的 Gradle 基础配置文件，用来配置项目的构建任务。代码如下所示。

```
1.  buildscript {
2.      //构建过程依赖的仓库
```

```
3.      repositories {
4.          //代码托管仓库
5.          jcenter()
6.      }
7.      dependencies {
8.          //Gradle 插件及使用版本
9.          classpath 'com.android.tools.build:gradle:3.2.0'
10.         //NOTE:Do not place your application dependencies here;they belong
11.         //in the individual module build.gradle files
12.     }
13. }
14. //这里面配置整个项目依赖的仓库,这样每个 module 就不用配置仓库了
15. allprojects {
16.     repositories {
17.         //代码托管仓库,可以引用 jcenter()上的任何开源项目
18.         jcenter()
19.     }
20. }
21. //运行 gradle clean 时,执行此处定义的 task
22. //该任务继承自 Delete,删除根目录中的 build 目录
23. //相当于执行 Delete.delete(rootProject.buildDir)
24. task clean(type: Delete) {
25.     delete rootProject.buildDir
26. }
```

app 下面的 build.gradle 文件主要配置应用属性、应用签名、应用特性(渠道)、应用构建类型和应用依赖。代码如下所示。

```
1.  apply plugin: 'com.android.application' //表示是一个应用程序的模块,可独立运行
2.  android {
3.      compileSdkVersion 28 //指定项目的编译版本
4.      defaultConfig {
5.          applicationId "cn.edu.jsit.smartfactory"   //指定包名
6.          minSdkVersion 17 //指定最低的兼容的 Android 系统版本
7.          targetSdkVersion 28 //指定的目标版本,表示在该版本的 Android 系统中已
经做过充分的测试
8.          versionCode 1          //版本号
9.          versionName "1.0"      //版本名称
10.         testInstrumentationRunner "android.support.test.runner.AndroidJUni
tRunner"
11.     }
12.     buildTypes {  //指定生成安装文件的配置
13.         release { //用于指定生成正式版安装文件的配置
14.             minifyEnabled false   //指定是否对代码进行混淆
15.             //指定混淆时使用的规则文件
16.             proguardFiles getDefaultProguardFile('proguard-android.txt'),
'proguard-rules.pro'
17.         }
18.     }
```

```
    19. }
    20. dependencies {  //指定当前项目的所有依赖关系：本地依赖、库依赖、远程依赖
    21.     implementation fileTree(dir: 'libs', include: ['*.jar']) //本地依赖
    22.     implementation 'com.android.support.constraint:constraint-layout:
1.1.3'
    23.     testImplementation 'junit:junit:4.12'
    24.     androidTestImplementation 'com.android.support.test:runner:1.0.2'
    25.     androidTestImplementation 'com.android.support.test.espresso:espress
o-core:3.0.2'
    26. }
```

7．工程目录中的关键文件和文件夹

下面是一些工程目录中的关键文件和文件夹，如图 2-12 所示。

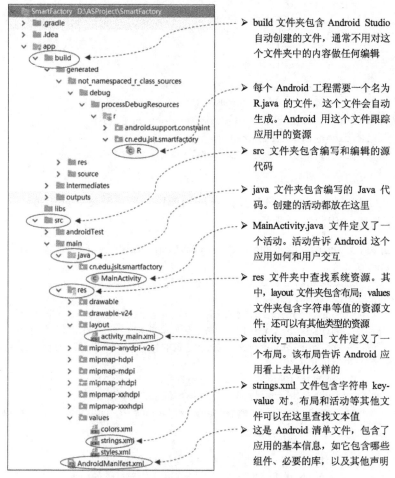

图 2-12　关键文件和文件夹

8．创建布局

双击 activity_main.xml 文件，可以开始编辑布局。布局编辑有两种方式，一种是通过代码编辑器进行编辑，另一种是通过设计编辑器进行编辑。代码编辑器就是一个文本编辑器，在编辑器中部分关键字会用不同的颜色进

7　创建布局和活动

行显示，如图 2-13 所示。

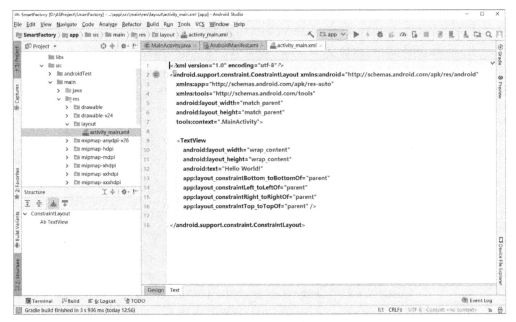

图 2-13　代码编辑器

单击编辑器左下角的"Design"按钮，进入设计编辑器，如图 2-14 所示。利用设计编辑器可以把 GUI 组件拖到布局中，根据设计的布局进行摆放。在右侧属性窗口中输入属性值，可以直接看到界面的效果，完成布局的设计。代码编辑器和设计编辑器是对同一个文件进行的不同视图下的编辑，两者之间可以来回切换。

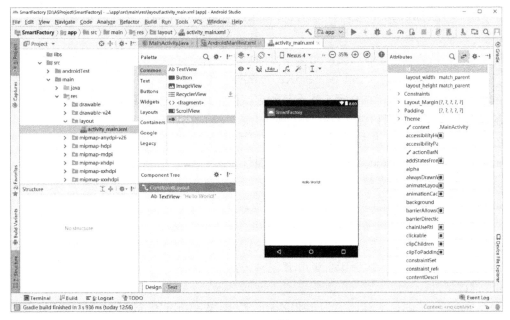

图 2-14　设计编辑器

下面看看这个简单布局文件的内容。

```
1.  <?xml version="1.0" encoding="utf-8"?>
2.  <android.support.constraint.ConstraintLayout
3.    xmlns:android="http://schemas.android.com/apk/res/android"
4.    xmlns:app="http://schemas.android.com/apk/res-auto"
5.    xmlns:tools="http://schemas.android.com/tools"
6.    android:layout_width="match_parent"
7.    android:layout_height="match_parent"
8.    tools:context=".MainActivity">
9.    <TextView
10.       android:layout_width="wrap_content"
11.       android:layout_height="wrap_content"
12.       android:text="Hello World!"
13.       app:layout_constraintBottom_toBottomOf="parent"
14.       app:layout_constraintLeft_toLeftOf="parent"
15.       app:layout_constraintRight_toRightOf="parent"
16.       app:layout_constraintTop_toTopOf="parent"/>
17. </android.support.constraint.ConstraintLayout>
```

第 1 行代码告诉解析器和浏览器，这个文件应该按照 1.0 版本的 XML 规则进行解析。ecoding="utf-8"表示此 XML 文件采用 UTF-8 的编码格式。

第 2 行声明布局采用的布局方式为 ConstraintLayout（约束布局）。ConstraintLayout 主要是为了解决布局嵌套过多的问题，以灵活的方式定位和调整小部件，从 Android Studio 2.3 起，官方的模板默认使用 ConstraintLayout。Android 的其他布局方式还有 RelativeLayout（相对布局）、LinearLayout（线性布局）、FrameLayout（帧布局）等，在后续章节中会进行介绍。

第 3～5 行是 Android 命名空间定义。和 Java 中的 package、C#中的 namespace 一样，这里的 XML 中的命名空间（xml namespace，xmlns）也是为了解决 XML 中元素和属性命名冲突的问题。因为 XML 中的标签并不是预定义的，这一点与 HTML 是有区别的，HTML 中的标签是预定义的，所以在 XML 中会遇到命名冲突的问题。

在 Android 中，目前遇到的 xmlns 一共有以下三种。

android：用于 Android 系统定义的一些属性。

app：用于应用自定义的一些属性。

tools：用于 XML 中的错误处理、预览和资源压缩等。

第 6、7 行让布局和设备的屏幕大小有相同的宽度和高度。

第 8 行表明 activity_main.xml 文件在 MainActivity.java 的 MainActivity 类里面有引用。

第 9～16 行定义了一个 TextView GUI 组件来显示文本。

9．创建活动

双击 MainActivity 文件，可以编辑活动，它由 Android Studio 自动生成，如图 2-15 所示。

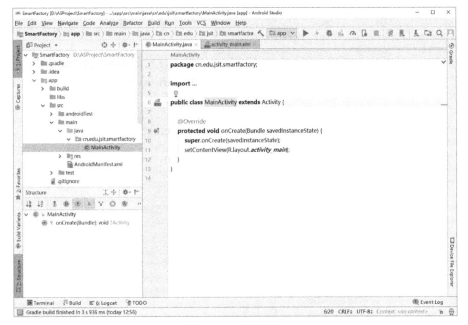

图 2-15 编辑活动

活动 MainActivity 就是一个 Java 类。

```
1.  package cn.edu.jsit.smartfactory;
2.  import android.app.Activity;
3.  import android.os.Bundle;
4.  public class MainActivity extends Activity {
5.      @Override
6.      protected void onCreate(Bundle savedInstanceState) {
7.          super.onCreate(savedInstanceState);
8.          setContentView(R.layout.activity_main);
9.      }
10. }
```

第 1 行声明了包名为 cn.edu.jsit.smartfactory。
第 2、3 行引入了 MainActivity 中使用的 Android 类。
第 4 行声明了 MainActivity 类继承了 Activity 类。
第 5~9 行重写了 Activity 类的 onCreate()方法。这个方法将在第一次创建活动的时候被调用。
第 7 行调用了父类的 onCreate()方法。
第 8 行指定要使用哪个布局文件。

2.4 运行应用

8 运行应用

运行应用有两个选择。可以选择在一个真正的物理设备上运行，如果没有 Android 手机或者平板电脑，可以选择在 Android 模拟器上运行。Android SDK 中内置了 Android 模拟器，这个模拟器允许建立一个或者多个 Android 虚拟设备（Android Virtual Device，AVD）。

在虚拟设备上运行应用与在真实的物理设备上运行一样，都可以测试应用。

在虚拟设备上运行应用的具体操作步骤如下。

1）在 Android Studio 中的模拟器上运行应用之前，首先要创建 Android 虚拟设备。在"Tools"（工具）菜单下，选择"AVD Manager"命令，进入 AVD 管理器，如图 2-16 所示。

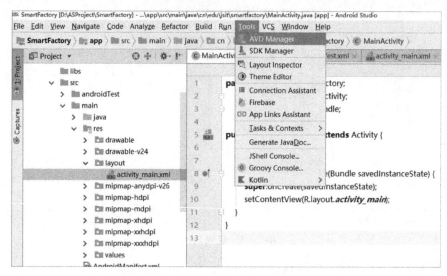

图 2-16　选择 AVD 管理器

2）要在 Android Studio 中创建一个 AVD，需要完成几个步骤。首先单击"Create Virtual Device"按钮，如图 2-17 所示。

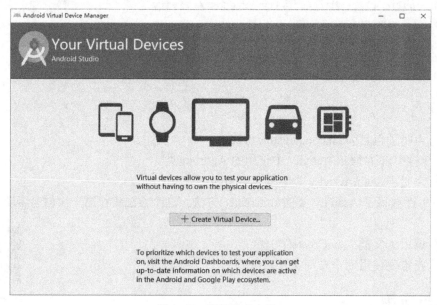

图 2-17　虚拟设备向导

3）这里需要选择一个将要模拟的硬件设备。可以选择不同的设备，包括 TV、手机、可穿戴设备、平板电脑。选择"Nexus 5X"这个已经存在的设备，在右边可以看到这个设备的一些物理参数（屏幕尺寸、分辨率），如图 2-18 所示。

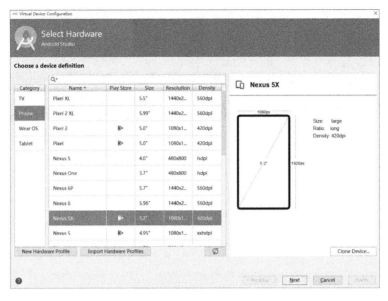

图 2-18 选择设备

4)如果在列表中没有满足需要的设备,可以单击左下角的"New Hardware Profile"按钮,自定义设备。单击"Next"按钮。

5)接下来,需要选择一个系统映像(system image)。系统映像提供了一个安装版本的 Android 操作系统,针对所构建的应用,要为它与兼容的 API 层次选择一个系统映像。如果希望应用在 API 17 以上运行,就要选择相应的系统映像(对应 API 17 以上)。这里选择 API 28 对象的系统映像,即"Pie 28 x86",目标为 Android 9.0,然后单击"Next"按钮。系统映像如果没有安装,Android Studio 会自动下载,可能需要较长的时间,如图 2-19 所示。

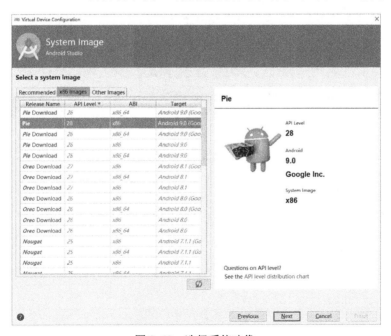

图 2-19 选择系统映像

25

6）接着验证 AVD 配置，在这个界面中会总结显示前几步选择的选项，并且允许进行修改，如图 2-20 所示。

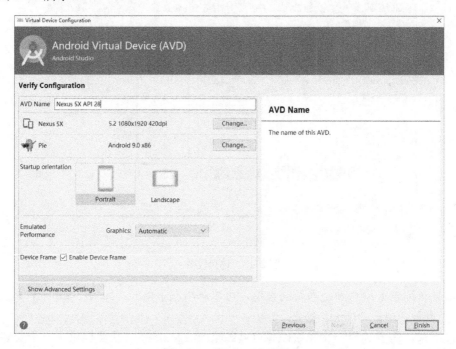

图 2-20　验证 AVD 配置

7）单击"Finish"按钮，完成 AVD 的创建。这时候在打开的虚拟设备列表窗口中会出现创建完成的虚拟设备，如图 2-21 所示。

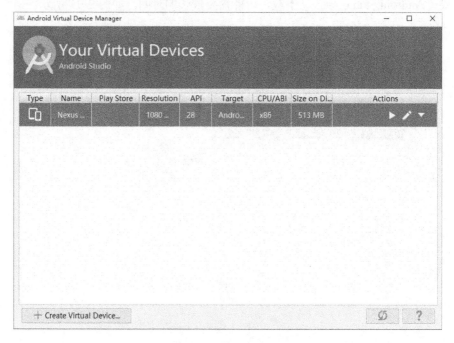

图 2-21　虚拟设备列表

8）接下来要让应用在创建的 AVD 上运行。在"Run"菜单下，选择"Run 'app'"命令，如图 2-22 所示。

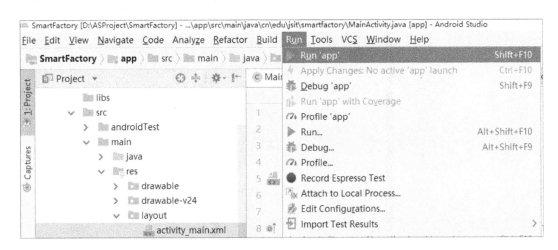

图 2-22　选择运行应用

9）选择部署的设备，当前可用的设备就是前面创建的"Nexus 5X API 28"。单击"OK"按钮，启动虚拟设备运行应用。勾选"Use same selection for future launches"复选框，下次该界面会自动跳过，默认选择这次所选的设备，如图 2-23 所示。

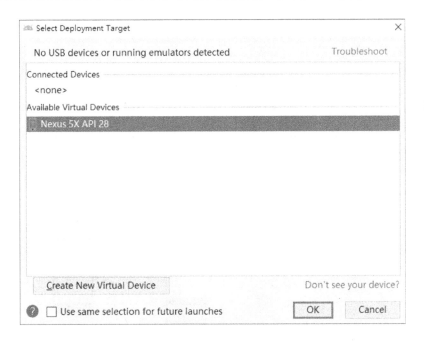

图 2-23　选择运行设备

系统启动后，自动安装运行 SmartFactory 应用，如图 2-24 所示。

图 2-24 应用运行界面

选择运行应用时并不是只运行 SmartFactory 这个应用，它还会运行应用所需的所有预备任务。下面说明应用编译、打包、部署和运行的过程，如图 2-25 所示。

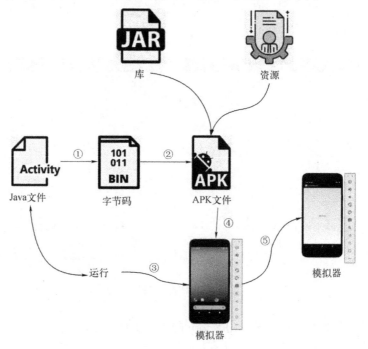

图 2-25 运行过程

① Java 源文件编译为字节码。

② 创建一个 Android 应用包或 APK 文件。APK 文件包含编译的 Java 文件以及应用所需的库和资源。

③ 如果模拟器还未运行，会启动模拟器并加载。

④ 一旦启动模拟器并加载 AVD，会将 APK 文件上传到 AVD 并安装。

⑤ AVD 启动应用关联的主活动。AVD 屏幕上会显示应用，可以开始测试。

可以在控制台上查看应用运行的进度。启动 AVD 通常需要几分钟，可以使用 Android Studio 控制台查看应用运行的过程，控制台详细记录了 Gradle 构建系统的过程，如果发生错误，可以通过控制台查看具体情况，如图 2-26 所示。

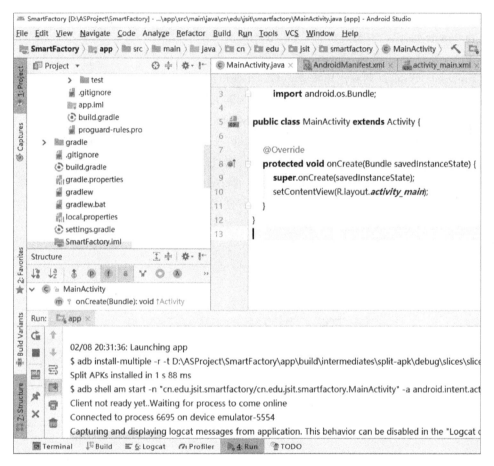

图 2-26　从控制台查看进度

从控制台可以看到过程如下。

① Android Studio 启动模拟器，加载 AVD，并安装应用。

② 应用启动时，从 MainActivity.java 创建一个活动。

③ 这个活动指定它要使用布局 activity_main.xml。

④ 活动告诉 Android 在屏幕上显示布局。另外会显示文字"Hello World!"。

2.5 更改应用的启动图标和应用名称

9 更改应用图标和名称

Android Studio 创建项目时会为应用自动创建应用的启动图标，放在 mipmap*文件夹内。在 Android 中，图片是资源的一种，一般应放在 app/src/main/res/drawable、drawable-hdpi、drawable-mdpi、drawable-xhdpi、drawable-xxhdpi、drawable-xxxhdpi 文件夹内。要在文件夹中增加一个图片文件，只需要把图片文件拖入这个文件夹。如果愿意，可以根据设备的屏幕密度使用不同的图片文件，这样可以在高密度屏幕上显示高分辨率图片，而在低密度屏幕上显示低分辨率图像。为此要在 app/src/main/res 中为不同的屏幕密度创建不同的 drawable 文件夹。文件夹与设备的屏幕密度相关，如下所示。

① drawable-mdpi：中密度屏幕，约 160dpi。
② drawable-hdpi：高密度屏幕，约 240dpi。
③ drawable-xhdpi：超高密度屏幕，约 320dpi。
④ drawable-xxhdpi：超超高密度屏幕，约 480dpi。
⑤ drawable-xxxhdpi：极高密度屏幕，约 640dpi。

然后将不同分辨率的图片放在各个 drawable*文件夹内，确保各个图片文件名字相同。Android 会根据运行设备的屏幕密度来确定运行时使用哪一个图片文件。举例来说，如果设备有一个超高密度屏幕，它就会使用位于 drawable-xhdpi 文件夹中的图片文件。如果只在一个文件夹中增加图片，Android 就会为所有设备使用相同的图片文件。如果这样，通常会将图片文件放在 drawable 文件夹中。图片资源目录如图 2-27 所示。

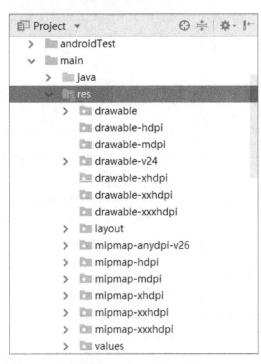

图 2-27 图片资源目录

这里注意，Android Studio 会自动生成 mipmap*文件夹。但一般在 mipmap*文件夹下，仅建议存放启动图标（app/launcher icons）和缩放动画相关的图片，而其他的图片资源等还是存放在 drawable*文件夹下。

因此，可以将应用的启动图标文件复制到 mipmap*文件夹内。将不同分辨率的 icon_launcher.png 图片复制到相应的 mipmap*文件夹内，如图 2-28 所示。

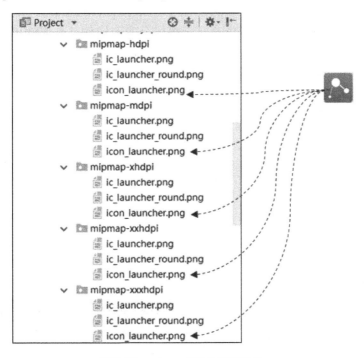

图 2-28　mipmap*图片资源目录

双击 AndroidManifest.xml 文件，修改相应属性。

```xml
1. <?xml version="1.0" encoding="utf-8"?>
2. <manifest xmlns:android="http://schemas.android.com/apk/res/android"
3.     package="cn.edu.jsit.smartfactory">
4.     <application
5.         android:allowBackup="true"
6.         android:icon="@mipmap/icon_launcher"
7.         android:label="@string/app_name"
8.         android:supportsRtl="true"
9.         android:theme="@style/AppTheme">
10.         <activity android:name=".MainActivity">
11.             <intent-filter>
12.                 <action android:name="android.intent.action.MAIN"/>
13.                 <category android:name="android.intent.category.LAUNCHER"/>
14.             </intent-filter>
15.         </activity>
16.     </application>
17. </manifest>
```

在第 6 行中，将 icon 修改为@mipmap/icon_launcher，重新运行应用。

AndroidManifest.xml 清单文件是每个 Android 程序中必需的文件。它位于整个项目的根目录，描述了 package 中各类组件（如 Activity、Service 等）、它们各自的实现类、各种能被处理的数据和启动位置。除了能声明程序中的各类组件，还能指定安全控制和测试。

第 2 行定义 Android 命名空间。

第 3 行指定本应用内 Java 主程序包的包名为"cn.edu.jsit.smartfactory"。

第 4～16 行在 application 标签中声明了每一个应用组件及其属性（如 icon、label、permission 等），这个标签不能少。

第 5 行定义允许 adb 进行备份和还原。

第 6 行定义应用图标。

第 7 行定义应用名称。

第 8 行定义支持 RTL。

第 9 行定义应用使用的主题。

第 10～15 行定义了活动 MainActivity。

第 10 行定义了活动的类名，有一个"."前缀，在这里这个类名就是.MainActivity。类名前面之所以有一个"."前缀，这是因为会结合类名和包名来得出完全限定类名"cn.jsit.edu.smartfactory.MainActivity"。

第 11～14 行定义了意图过滤器，指定了应用启动时加载 MainActivity 这个活动。

更新后运行结果如图 2-29 所示。

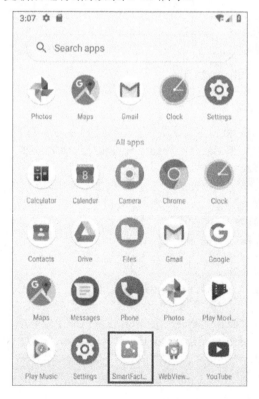

图 2-29　更新启动图标

打开 app/src/main/res/values 目录下的 strings.xml 文件。Android Studio 创建了一个字符串资源文件，名为 strings.xml，通过 name-value 来存放字符串，在布局文件中包含字符串的引用，而不是字符串本身。

```
1. <resources>
2.     <string name="app_name">SmartFactory</string>
3. </resources>
```

将 app_name 改为"智慧工厂"，如图 2-30 所示。

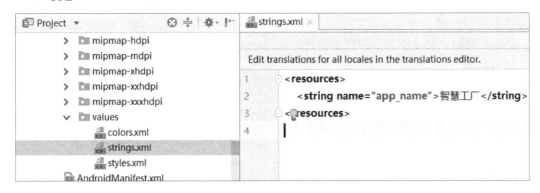

图 2-30　在 strings.xml 中更改应用名称

重新运行程序，结果如图 2-31 所示。

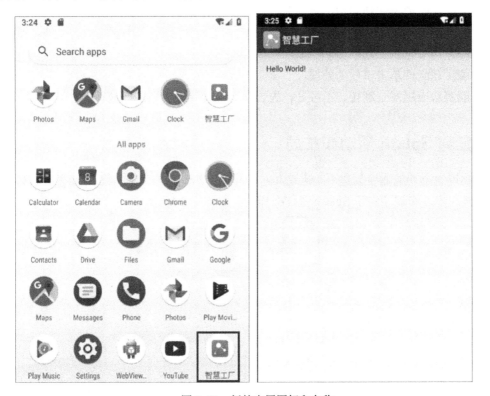

图 2-31　新的应用图标和名称

任务 3　创建 Splash 界面

任务概述

任务 2 已经构建了一个基本的 Android 应用，而且查看了它在模拟器中的运行情况。本任务要为应用创建一个 Splash 界面。Splash 界面是应用启动界面，停留 6s 后进入应用主界面。Splash 界面主要用于显示应用的信息。

知识目标

- 掌握 TextView 文本视图组件。
- 了解 RelativeLayout（相对布局）。
- 了解 Handler（消息处理器）调度代码方法。
- 掌握 Intent（意图）概念。
- 理解 R.java 的作用。
- 掌握 AndroidManifest 文件的作用。
- 理解活动的生命周期。

技能目标

- 能创建、编辑活动和布局。
- 能使用 Intent 实现活动跳转。
- 能使用 Handler 调度运行代码。
- 能实现组件单击事件的处理。
- 能解决应用发生旋转、不可见、失去焦点等情况下出现的问题。

3.1　创建 Splash 活动和布局

要实现这个功能需要创建一个活动和一个布局。活动指定了 Splash 界面做什么以及应当如何响应用户，布局指定了 Splash 是什么样的，如图 3-1 所示。

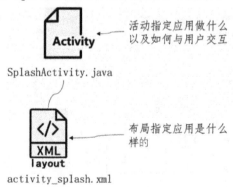

图 3-1　活动与布局

新增一个活动，可以打开前面创建的 SmartFactory 项目，右击并在弹出的快捷菜单中选择"New"→"Activity"→"Empty Activity"命令，如图 3-2 所示。

图 3-2　新增活动

下面的步骤前面已经讲过，为活动指定名称为"SplashActivity"，勾选"Generate Layout File"复选框自动生成布局文件，指定布局名称为"activity_splash"，如图 3-3 所示。单击"Finish"按钮。

图 3-3　指定活动和布局名称

Android Studio 自动生成了 SplashActivity.java 活动和 activity_splash.xml 布局，如图 3-4 所示。

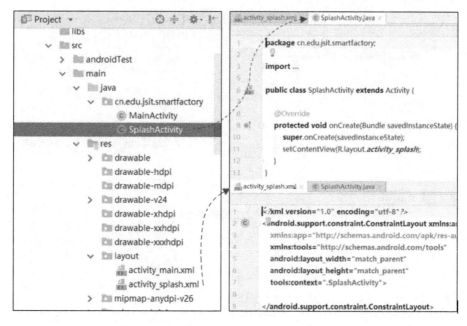

图 3-4　Splash 活动和布局

在 AndroidManifest.xml 清单文件中增加了一行，添加了 SplashActivity 活动。

```
<activity android:name=".MainActivity"></activity>
```

3.2　编辑 Splash 布局

向布局增加 GUI 组件有两种方法，一种是直接在布局文件中增加，另一种是使用设计编辑器增加。下面通过设计编辑器增加一个文本视图。

10　编辑 Splash 布局

在设计编辑器左边有一个组件面板，其中有很多 GUI 组件，可以将其直接拖到设计编辑器中。如果查看组件面板中的 Text 栏，可以看到其中有一个 TextView 组件，单击这个组件，把它拖到设计编辑器中，如图 3-5 所示。

设计编辑器中的修改会反映到 XML 文件中，像这样把 GUI 组件拖到设计编辑器中可以很方便地更新布局。如果切换到代码编辑器，可以看到，通过设计编辑器增加文本视图组件，也会相应在文件中增加几行代码。

```
1. <TextView
2.     android:id="@+id/textView"
3.     android:layout_width="wrap_content"
4.     android:layout_height="wrap_content"
5.     android:layout_alignParentTop="true"
6.     android:layout_centerHorizontal="true"
```

7.   android:layout_marginTop="176dp"
8.   android:text="TextView"/>

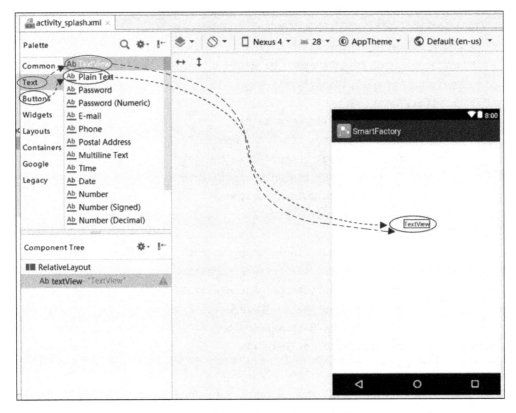

图 3-5 在设计编辑器添加 GUI 组件

一个 GUI 组件有多个属性，一些属性是共有的。例如以下几个。

android:id，这个属性为组件指定一个标识名。利用 id 属性，可以通过代码来控制组件完成工作，另外还可以控制组件在布局中的位置。

    android:id="@+id/textView"

android:text，告诉 Android 组件应当显示什么文本。

    android:text="TextView"

android:layout_width，android:layout_height，这两个属性指定了组件的基本宽度和高度（是不能缺少的）。参数 wrap_content 表示它要足够大，刚好能放下文本的全部内容。

    android:layout_width="wrap_content"
    android:layout_height="wrap_content"

要在启动界面中增加两个文本视图，一个是"欢迎使用智慧工厂应用"，另一个是"版本 1.0"，两个文本居中显示在屏幕中间，上下排列。下面仔细分析一下布局代码，以便了解其究竟在做什么（如果你的代码看上去有些不同也没有关系，请跟着一步一步往下走即可)。进入 app/src/main/res/layout 文件夹并双击打开 activity_splash.xml 文件。

```xml
1.  <?xml version="1.0" encoding="utf-8"?>
2.  <RelativeLayout xmlns:android="http://schemas.android.com/apk/res/android"
3.  xmlns:tools="http://schemas.android.com/tools"
4.  android:layout_width="match_parent"
5.  android:layout_height="match_parent"
6.  tools:context=".SplashActivity">
7.      <TextView
8.          android:id="@+id/welcome_tv"
9.          android:layout_width="wrap_content"
10.         android:layout_height="wrap_content"
11.         android:layout_centerInParent="true"
12.         android:text="@string/welcome"
13.         android:textSize="28sp"/>
14.     <TextView
15.         android:layout_width="wrap_content"
16.         android:layout_height="wrap_content"
17.         android:layout_centerInParent="true"
18.         android:layout_below="@+id/welcome_tv"
19.         android:text="@string/version"
20.         android:textSize="16sp"/>
21. </RelativeLayout>
```

布局代码中的第一个元素是<RelativeLayout>。<RelativeLayout>元素告诉 Android 布局中的各个 GUI 组件要在相对位置上显示，这种布局方式称为相对布局，它是 Android 的一种重要的布局方式。相对布局，意思是指该布局中的所有 GUI 组件之间的位置关系是相对放置的。例如，可以用这个元素指出你希望某一个组件出现在另一个组件的右边，或者希望这些组件以某种方式对齐或排列。

第 7～13 行是第一个 TextView 文本视图组件，14～20 行是第二个文本视图组件。

第 8 行定义了第一个文本视图的 id 为 "welcome_tv"。

第 9、10 行指定了宽度和高度。

第 11 行指定了将文本在父布局中垂直和水平居中。

第 12 行通过引用字符串资源文件 strings.xml 中名为 welcome_tv 的文本值，指定文本显示的内容。这里需要先在 string.xml 文件中增加一个字符串资源。

```xml
<string name="welcome_message">欢迎使用智慧工厂应用</string>
```

第 13 行指定了文本显示大小，以 sp 为单位。sp 是缩放无关的抽象像素（scale-independent pixel, sp），其作为字体大小单位会随着系统的字体大小改变，建议在设置字体大小时使用 sp 作为单位。

第 14～20 行定义了第二个文本视图。

第 15、16 行指定了宽度和高度。

第 17 行指定了将文本在父布局中垂直和水平居中。

第 18 行指定了将该文本视图放置在 welcome_tv 文本视图下面。

第 19 行通过引用字符串资源文件 strings.xml 中名为 version 的文本值，指定文本显示的内容。这里需要先在 string.xml 文件中增加一个字符串资源。

```
<string name="version">版本 1.0</string>
```

在设计编辑器中可以看到预览界面，如图 3-6 所示。

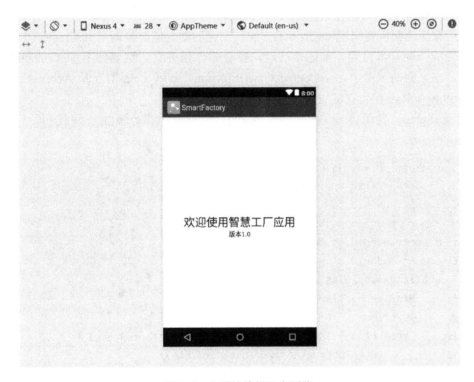

图 3-6 在设计编辑器中预览

3.3 编辑 Splash 活动

通过向导完成了 SplashActivity 活动的创建，但到目前为止还什么都没做，我们希望这个活动能够使 Splash 界面停留 6s 后再进入主界面。下面分析一下活动代码，进入 app/src/main/java 文件夹并双击打开 SplashActivity.java 文件。

11 编辑 Splash 活动

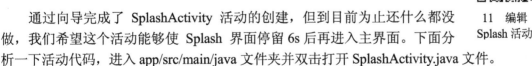

```
6.    protected void onCreate(Bundle savedInstanceState) {
7.        super.onCreate(savedInstanceState);
8.        setContentView(R.layout.activity_splash);
9.    }
10. }
```

以上就是创建一个基本活动所需的全部代码。可以看到，这是一个继承了 android.app.Activity 的类，而且实现了一个 onCreate()方法。所有活动都必须继承 Activity 类。Activity 类包含一组方法，正是这组方法为 Java 类赋予了 Android 生命，把它从一个普通的 Java 类变成了一个功能完备的正式的 Android 活动。

另外所有活动都需要实现 onCreate()方法。创建活动对象时会调用 onCreate()方法，这个方法用来完成一些基本设置，如这个活动与哪个部件相关联。这要使用 setContentView()方法来设置。在上面的示例中，setContentView(R.layout.activity_splash)告诉 Android 这个活动会使用 activity_splash 作为它的布局。

为实现 Splash 界面停留 6s 再进入主界面，需要为活动添加一些代码。下面更新 SplashActivity.java。

```
1.  package cn.edu.jsit.smartfactory;
2.  import android.app.Activity;
3.  import android.content.Intent;
4.  import android.os.Bundle;
5.  import android.os.Handler;
6.  import android.os.Message;
7.  public class SplashActivity extends Activity {
8.      Handler handler=new Handler();
9.      @Override
10.     protected void onCreate(Bundle savedInstanceState) {
11.         super.onCreate(savedInstanceState);
12.         setContentView(R.layout.activity_splash);
13.         goToMain();
14.     }
15.     private void goToMain() {
16.         handler.postDelayed(new Runnable() {
17.             public void run() {
18.                 Intent intent=new Intent(SplashActivity.this, MainActivity.class);
19.                 startActivity(intent);
20.             }
21.         }, 6000);
22.     }
23. }
```

第 8 行是新增代码，定义了一个 Handler 对象。Handler（消息处理器）是一个 Android

类，可以用来调度要在将来某个时刻运行的代码，还可以用它提交在不同线程中运行的代码。这里需要调出在 6s 后跳转的主界面。在 goToMain()方法中，使用 postDelayed()方法实现。

使用 Handler 时，可以把想要调度的代码包装在一个 Runnable 对象中，然后使用 Handler 的 post()和 postDelayed()方法指定希望这个代码在什么时候运行。

post()方法提交的代码需要尽可能快地运行（通常几乎是立即运行）。post()方法有一个参数，这是一个类型为 Runnable 的对象。Android 里的 Runnable 对象与普通 Java 中的 Runnable 对象很相似，就是要运行的一个作业。可以把想要运行的代码放在 Runnable 对象的 run()方法中，Handler 会确保这个代码尽可能快地运行。具体方法如下。

```
final Handler handler=new Handler();
handler.post(Runnable);
```

postDelayed()方法的工作与 Post()方法类似，不过这个方法用来提交将来运行的代码。postDelayed()方法有两个参数，一个是 Runnable 对象，另一个是 long。运行的代码包含在 Run()方法中。long 参数指定了希望代码延迟多少毫秒运行，在这个延时后，代码会尽可能快地运行。具体方法如下。

```
final Handler handler=new Handler();
handler.postDelaged(Runnable,long);    //使用这个方法让代码延迟指定的时间（毫秒数）
                                       //后再运行
```

这里使用了 postDelayed()方法，long=6000（即 6s）。

在 run()方法中有两行代码，实现了跳转的功能。

```
Intent intent=new Intent(SplashActivity.this, MainActivity.class);
startActivity(intent);
```

需要让一个活动启动另一个活动，需要使用一个意图（Intent）。可以把 Intent 看成是一个"想要做某事情的意图"。这个消息类型允许你在运行时把单独的对象（如活动）绑定在一起。如果一个活动想要启动第二个活动，可以向 Android 发送一个意图。Android 会启动第二个活动，并传入意图。只需几行代码就可以创建和发送一个意图。首先创建如下的意图。

```
Intent intent=new(this,Target.class);
```

第一个参数告诉 Android 这个活动来自哪个对象，可以用 this 表示当前活动。第二个参数是接收了这个意图的活动的类名。创建意图后，把它传入 Android，代码如下。

```
startActivity(intent);
```

这会告诉 Android 要启动这个意图指定的活动。Android 接收到这个意图后，会监测是否一切正常，并要求这个活动启动；如果无法找到这个活动，它会抛出一个 ActivityNotFoundException 异常。

本例中简单介绍了消息处理器（Handler）和意图（Intent），在后面我们还会遇到并详细介绍。

需要更新 AndroidManifest.xml 文件，将应用最先启动活动设置为 SplashActivity，如下所示。

```
1. <activity android:name=".SplashActivity">
2.     <intent-filter>
3.         <action android:name="android.intent.action.MAIN" />
4.         <category android:name="android.intent.category.LAUNCHER" />
5.     </intent-filter>
6. </activity>
```

第 3 行中 android.intent.action.MAIN 决定应用的入口 Activity，也就是启动应用时首先显示哪一个 Activity。

第 4 行中 android.intent.category.LAUNCHER 表示 Activity 应该被列入系统的启动器（Launcher），允许用户启动它。Launcher 是 Android 系统中的桌面启动器，是桌面 UI 的统称。

运行这个应用可以看到如图 3-7 所示的结果。

图 3-7　界面跳转

12　添加延时计时和取消

一般 Splash 界面右上角还会显示延时 "xs" 的描述和 "跳过"，用户可以单击 "跳过" 直接进入主界面，而不需要等待。这需要在布局文件中增加两个 TextVeiw 组件。

```
1. <TextView
2.     android:id="@+id/tv_count"
3.     android:layout_width="wrap_content"
4.     android:layout_height="wrap_content"
5.     android:layout_alignParentRight="true"
6.     android:layout_marginTop="10dp"
7.     android:layout_marginRight="50dp"
8.     android:text="6s"/>
9. <TextView
10.    android:id="@+id/tv_cancel"
```

```
11.     android:layout_width="wrap_content"
12.     android:layout_height="wrap_content"
13.     android:layout_alignParentRight="true"
14.     android:layout_marginTop="10dp"
15.     android:layout_marginRight="10dp"
16.     android:text="@string/cancel"
17.     android:onClick="onClickCancel"/>
```

第 5、13 行中，android:layout_alignParentRight 的作用是把视图与父视图右对齐。

第 6、14 行中，android:layout_marginTop 的作用是设置在视图上方额外增加的空间，单位是 dp。dp 是设备（密度）无关像素（density-independent pixels，dp），有些设备使用非常小的像素，能创建高清晰的图像。另外一些设备则比较廉价，因为这些设备的像素比较少，也比较大。通过使用设备（密度）无关像素，可以避免所创建的界面在某些设备上太小而在另外一些设备上又过大。如果按设备无关像素来度量，那么在所有设备上大小都几乎相同。

第 7、15 行中，android:layout_marginRight 的作用是设置在视图右边额外增加的空间，单位也是 dp，一般这些图像单位都使用 dp，如果是设置文字大小，则使用 sp 单位。

第 17 行中，android:onClick="onClickCancel"指定该文本视图被单击后要调用的方法名称为 onClickCancel。当用户单击"取消"时，调用活动 SplashActivity 中的 onClickCancel() 方法，通过这个方法实现跳过功能，直接进入主界面。

现在布局知道了要调用活动中的哪个方法，接下来要在活动中编写这个方法，同时还要对 goToMain()方法进行修改。

```
1.  package cn.edu.jsit.smartfactory;
2.  import android.app.Activity;
3.  import android.content.Intent;
4.  import android.os.Bundle;
5.  import android.os.Handler;
6.  import android.view.View;
7.  import android.widget.TextView;
8.  public class SplashActivity extends Activity {
9.      private int seconds=6;
10.     private boolean skipping=false;
11.     Handler handler=new Handler();
12.     @Override
13.     protected void onCreate(Bundle savedInstanceState) {
14.         super.onCreate(savedInstanceState);
15.         setContentView(R.layout.activity_splash);
16.         goToMain();
17.     }
18.     private void goToMain() {
19.         final TextView timeView=(TextView) findViewById(R.id.tv_count);
20.         handler.post(new Runnable() {
```

```
21.            @Override
22.            public void run() {
23.                String time=String.format("%d"+"s", seconds);
24.                timeView.setText(time);
25.                if((seconds==0)||(skipping==true)) {
26.                    Intent intent=new Intent(SplashActivity.this, MainActivity.class);
27.                    startActivity(intent);
28.                } else {
29.                    seconds--;
30.                    handler.postDelayed(this, 1000);
31.                }
32.            }
33.        });
34.    }
35.    public void onClickCancel(View view) {
36.        skipping=true;
37.    }
38. }
```

第 9 行定义了整型变量 seconds，记录定时剩余秒数，初始值为 6。

第 10 行定义了布尔变量 skipping，是判断是否"取消"的依据。

第 35～37 行定义了 onClickCancel()方法。这个方法的形式为"public void onClickCancel (View view)"，如果未采用这种形式，用户单击"取消"时，这个方法不会有任何响应。这是因为在后台 Android 会查找一个有 void 返回值的公共方法，而且方法名要与布局 XML 中指定的方法匹配。

方法中的 View 参数看起来很奇怪，不过这个参数的存在是有道理的。这个参数指示触发这个方法的 GUI 组件（在这里就是文本视图）。文本视图（TextView）是 View 的一种。

定义该方法要使用 View，所以在第 6 行导入了这个类。当用户单击"取消"时，Android 调用该方法，将 skipping 设为 ture，在 goToMain()方法中可以判断 skipping 的值来决定是否跳过延时。

第 19 行定义了一个 TextView 的对象，我们用一个名为 findViewById()的方法得到这个 GUI 组件的句柄。findViewById()方法取 GUI 组件的 id 作为参数，返回一个 View 对象，然后把这个返回值转换为正确的 GUI 组件类型（TextView）。在代码中：

```
TextView timeView=(TextView)findViewById(R.id.tv_count);
```

使用 findViewById()得到了 id 为 tv_count 的文本视图的一个应用。仔细查看如何指定文本视图的 ID，这里并不是传入文本视图的名字，而是传入了一个形如 R.id.tv_count 的 ID，这到底是什么意思呢？什么是 R？

R.java 是一个特殊的 Java 文件，只要创建或构建应用，Android 工具就会生成这个文件。它位于工程 app/build/generated/source/debug 文件夹下的一个包中，这个包与应用包同名。Android 使用 R 跟踪应用中使用的资源，它有很多作用，如允许在活动代码中得到 GUI 组件的引用。

如果打开 R.java，可以看到其中包含一系列内部类，分别对应不同类型的资源。可以在内部类中引用各种类型的资源。例如，R 中有一个名为 id 的内部类，这个内部类包含一个 static final tv_count。利用代码(TextView)findViewById(R.id.tv_count)，可以由这个值得到 tv_count 文本视图的引用。

第 20～33 行使用 Handler 的 post()方法提交代码。

第 23 行定义了 String 类型的变量 time（秒数值），通过 seconds 转换而来。

第 24 行文本视图对象 timeView 调用 setText()方法，设置该文本视图的文本为 time（秒数值）。

第 25 行判断当前延时的秒数值是否为 0，或者用户是否单击了"取消"使得 skipping 的值为 true，只要两个条件满足其一，就跳转到活动 MainActivity。

第 29、30 行中，如果前面两个情况都没出现，则 seconds 减 1，并且使用 Handler 的 postDelayed()方法提交代码，1s 后继续执行 Run()方法。

运行这个应用可以看到如图 3-8 所示的结果。

图 3-8 界面延时跳转

3.4 修改活动及其生命周期

当使用 Android 应用时，旋转设备的情况经常发生。这时会发现当进入主界面前如果旋转设备，延时计数会重新开始，如图 3-9 所示。这是怎么回事呢？

13 修改活动及其生命周期 1

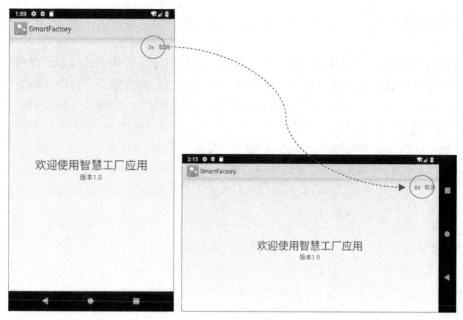

图 3-9 界面旋转

原因是 Android 看到屏幕方向和屏幕大小发生变化时，会撤销这个活动，然后重新创建 SplashActivity 活动，seconds 值被重新赋为 6，onCreate()方法会运行，然后再执行 goToMain()方法。

Android 会运行并启动一个活动，它会考虑设备的配置。这是指物理设备的配置（如屏幕大小、屏幕方向，以及是否有一个关联的键盘），另外还会考虑用户指定的配置选项（如本地化环境）。Android 启动活动时需要知道设备配置是什么，因为这可能会影响应用所需的资源。例如，如果设备屏幕是水平而不是垂直的，可能就需要一个不同的布局；另外如果本地化环境是英语环境，可能需要一组不同的字符串值。设备配置改变时，显示用户界面的所有组件都需要更新，从而与新配置一致。如果旋转了设备，Android 会发现屏幕方向和屏幕大小有变化，它会把这种情况归为设备配置的变化，所以会撤销当前活动，然后重新创建这个活动，以便选择适合新配置的资源。

Android 创建和撤销活动时，这个活动会从启动状态进入运行状态，然后再被撤销。活动的主状态就是运行或活动状态。如果一个活动在屏幕前台，而且得到了焦点，用户可以与之交互，这个活动就处于运行状态。活动的一生（生命周期，lifecycle）的大部分时间都处于这个状态，如图 3-10 所示。

活动从启动状态到撤销状态，会触发几个关键的活动生命周期方法：onCreate()和 onDestroy()。它们是活动继承得到的生命周期方法，如果需要，可以重写（override）这些方法。活动启动后会立即调用 onCreate()方法。活动的所有设置工作就在这个方法中完成，如调用 setContentView()方法。一般都要重写这个方法。如果没有重写，可能就无法告诉 Android 你的活动要使用哪个布局。onDestroy()方法是活动撤销前最后调用的方法。很多情况下有可能会撤销活动，如活

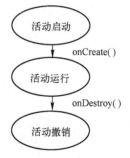

图 3-10 活动的生命周期

动完成，或者由于设备配置发送变化导致活动重建，也可能是 Android 决定撤销这个活动以节省空间。

在 SplashActivity 中，SplashActivity 类继承了 android.app.Activity 类。正是因为这个类，活动才可以访问 Android 生命周期方法，继承关系如图 3-11 所示。

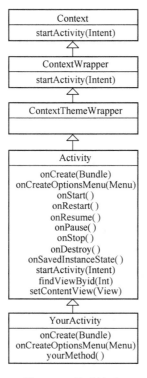

图 3-11 继承关系

Context(android.content.Context)是一个抽象类，是应用环境全局信息的一个接口。运行访问应用资源、类和应用级操作。ContextWrapper(android.content.ContextWrapper)是 Context 的一个代理实现。ContextThemeWrapper(android.view.ContextThemeWrapper) 允许修改 ContextWrapper 中的主题。Activity(android.app.Activity)类实现了生命周期方法的默认版本。它还定义了另外一些方法，如 findViewById(Int)和 setContentView(View)。YourAcitvity(cn.edu.jsit.smartfactory)是用户类，在用户类中活动的大多数行为都由父类方法处理，当需要时可以重写部分方法。

上面这种在旋转设备时出现的问题如何解决？也就是该如何处理配置变化，避开重建活动呢？有两种方法。

第一种方法是可以告诉 Android 不要重启活动，或者可以保存活动的当前状态，这样活动就能自行重新创建并恢复到原来的状态。

如果选择这种方法，可以在 AndroidManifest.xml 文件的 Activity 元素中增加一行，如下所示。

```
android:configChange="configration_change"
```

configuration_change 是配置改变的类型,通过设置告诉 Android 不要考虑屏幕方向和屏幕大小的变化。

```xml
1. <activity android:name=".SplashActivity"
2. android:configChanges="orientation|screenSize">
3.   <intent-filter>
4.     <action android:name="android.intent.action.MAIN"/>
5.     <category android:name="android.intent.category.LAUNCHER"/>
6.   </intent-filter>
7. </activity>
```

第 2 行说明要避开这两种配置变化。如果 Android 遇到这种配置变化,它会发出一个 onConfigrationChanged(Configuration)方法调用而不是重新创建活动,如果需要,可以实现这个方法对配置变化做出反应。

第二种方法是保存当前状态,当活动重建时,在 onCreate()中恢复当前保存的状态。要实现活动状态的保存,需要用到 onSaveInstanceState()方法,这个方法在活动撤销之前会被 Android 调用,可以利用它来保存某些值。

```java
1. @Override
2. public void onSaveInstanceState(Bundle savedInstanceState) {
3.     savedInstanceState.putInt("seconds",seconds);
4. }
```

onSaveInstanceState()方法有一个参数 Bundle,利用它的 put*("name", value)方法可以把不同类型的数据以键/值(Key-Value)对添加到一个对象中。当 SplashActivity 活动在配置变化被撤销时,该方法会被 Android 调用,将当前的 seconds 保存在 savedInstanceState 对象中。在 SplashActivity 的 onCreate()中可以再从该对象中得到 seconds 的值。

```java
1. @Override
2. protected void onCreate(Bundle savedInstanceState) {
3.     super.onCreate(savedInstanceState);
4.     setContentView(R.layout.activity_splash);
5.     if(savedInstanceState!=null) {
6.         seconds=savedInstanceState.getInt("seconds");
7.     }
8.     goToMain();
9. }
```

onCreate()方法有一个参数 Bundle,如果这个活动是第一次创建的,这个参数是 null;如果这个活动是重新创建的,之前已经调用过 onSaveInstanceState()方法,就会把 onSaveInstanceState()使用的 Bundle 对象传递给活动。

如果在 Splash 界面延时计数时,突然进来了一个电话,这时 Splash 界面就会变得不可见,SplashActivity 活动仍然在运行,等电话打完时,应用已经进入主界面了。如果希望电话进来时,SplashActivity 活动能停止运行,等电话结束后,SmartFactory 应用再次可见时,再恢复运行,这就需要用到活

14 修改活动及其生命周期2

动的其他一些生命周期方法。onCreate()方法和 onDestroy()方法可以处理活动的整个生命周期，另外还有 3 个关键方法可以用来处理活动对用户可见或不可见的工作，分别是 onStart()、onStop() 和 onRestart()，这些方法也是活动从 android.app.Activity 类继承的。

活动对用户可见时会调用 onStart()，不可见时会调用 onStop()。可见或不可见可能是因为活动被在它上面显示的另一个活动隐藏，也有可能是这个活动被撤销。如果活动被撤销而调用 onStop()，调用 onStop()之前会调用 onSaveInstanceState()。如果活动不可见，在它重新变为可见时，会调用 onRestart()。活动生命周期如图3-12所示。

活动启动并运行 onCreate()方法，初始化代码。此时活动还不可见，因为还没有调用 onStart()。onStart()方法运行后，用户将在屏幕上看到这个活动。当活动对用户不可见时，onStop()方法被调用，之后活动不再可见。如果活动再次可见时，onRestart()被调用，活动在不可见和可见之间来回切换会反复经过这个循环。撤销活动时会先调用 onStop()方法，然后调用 onDestroy()方法，如果设备内存很低，onStop()可能被跳过。

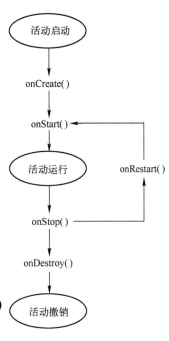

图3-12 活动生命周期

为了在电话结束后，让 SplashActivity 重新恢复运行，可以在 onStop()方法中让延时计数停止运行，等到恢复（再次获得焦点）继续运行时再在 onStart()方法中让延时计数继续运行。

```
1.  public class SplashActivity extends Activity {
2.      private int seconds=6;
3.      private boolean skipping=false;
4.      Handler handler=new Handler();
5.      private boolean running=true;
6.      private boolean wasRunning;
7.
8.      @Override
9.      protected void onCreate(Bundle savedInstanceState) {
10.         super.onCreate(savedInstanceState);
11.         setContentView(R.layout.activity_splash);
12.         if(savedInstanceState!=null) {
13.             seconds=savedInstanceState.getInt("seconds");
14.             wasRunning=savedInstanceState.getBoolean("wasRunning");
15.         }
16.         goToMain();
17.     }
18.
19.     private void goToMain() {
20.         final TextView timeView=(TextView) findViewById(R.id.tv_count);
```

```
21.         handler.post(new Runnable() {
22.             @Override
23.             public void run() {
24.                 String time=String.format("%d"+"s", seconds);
25.                 timeView.setText(time);
26.                 if((seconds==0)||(skipping==true)) {
27.                     Intent intent=new Intent(SplashActivity.this, MainActivity.class);
28.                     startActivity(intent);
29.                 } else {
30.                     if(running) {
31.                         seconds--;
32.                     }
33.                     handler.postDelayed(this, 1000);
34.                 }
35.             }
36.         });
37.     }
38.
39.     public void onClickCancel(View view) {
40.         skipping=true;
41.     }
42.
43.     @Override
44.     public void onSaveInstanceState(Bundle savedInstanceState) {
45.         savedInstanceState.putInt("seconds",seconds);
46.         savedInstanceState.putBoolean("wasRunning",wasRunning);
47.     }
48.
49.     @Override
50.     protected void onStop() {
51.         super.onStop();
52.         wasRunning=running;
53.         running=false;
54.     }
55.
56.     @Override
57.     protected void onStart() {
58.         super.onStart();
59.         if(wasRunning) {
60.             running=true;
61.         }
62.     }
63. }
```

第 5 行定义了一个 running 的变量，记录当前延时计数是否在运行，初始状态为 true。

第 6 行定义了一个 wasRunning 的变量，记录调用 onStop()方法之前延时计数是否在运行，这样就能知道活动再次可见时是否让它再次运行。

第 14 行当活动重新创建时，恢复 wasRunning 变量的状态。

第 46 行保存 wasRunning 变量的状态。

第 49～54 行重写了 onStop()方法。

第 51 行调用父类 onStop()方法。如果重写一个 Android 生命周期方法，必须调用父类的生命周期方法，这里是 super.onStop()。因为首先要确保这个活动完成父类生命周期方法中的所有动作，如果跳过这一步，Android 会抛出一个异常。

第 52 行将当前 running 状态赋给 wasRunning。

第 53 行将当前 running 状态设置为 false。

第 56～62 行重写了 onStart()方法。

第 58 行调用父类 onStart()方法。

第 59 行重建活动时判断原来运行状态，如果为真，则第 60 行中当前 running 的状态设置为 true（继续延时计数）。

还有一种情况，当 SplashActivity 活动运行时，另外一个活动出现在它之上，使得 SplashActivity 失去了焦点，但是这时 SplashActivity 活动还是可见的，这时 SplashActivity 会进入暂停状态，SplashActivity 活动仍然存在，而且会维护它的所有状态信息，如图 3-13 所示。

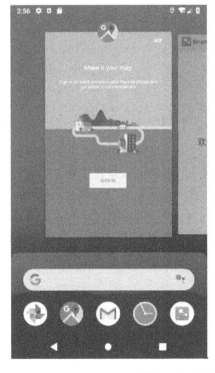

图 3-13　失去焦点活动暂停

如果活动可见，但是没有得到焦点（见图 3-13 左图），活动会暂停。如果得到焦点，活动会继续运行（见图 3-13 右图）。有两个生命周期方法可以用来处理活动的暂停和活动再次启动，分别是 onPause()和 onResume()。如果出现左图情况，SplashActivity 就会调用 onPause()方法，当 SplashActivity 再次获得焦点时，会调用 onResume()方法。如果希望暂停时 SplashActivity 活动要做一些处理，可以实现两个方法。活动生命周期如图 3-14 所示。

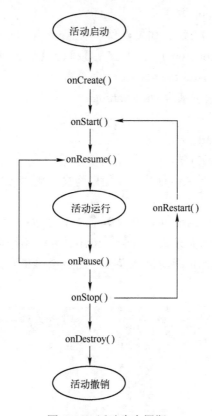

图 3-14 活动生命周期

当活动启动时，运行 onCreate()和 onStart()方法，这时活动是可见的，但是还没有得到焦点。onResume()方法运行之后，活动得到了焦点，用户可以与它交互。当活动失去焦点时（活动仍可见），onPause()方法被调用。如果活动再次进入前台获得焦点，onResume()会被调用。活动不可见时，onStop()会被调用，活动再次可见时，会调用 onRestart()方法，接下来会调用 onStart()和 onResume()方法。活动从运行状态到撤销状态时，在活动撤销前会调用 onPause()方法，通常还会调用 onStop()方法。

任务 4　创建系统主界面

任务概述

主界面中要向用户显示的内容有各类传感器的信息，关键非工作区域（危险区域）是否有人闯入的信息，还有通风系统和光照系统的开启、关闭、自动调节控制。主界面可以分为 3 个区域，上面的区域显示环境监控信息，主要显示温度传感器、湿度传感器、光照传感器的信息，中间区域显示非工作区域是否有人闯入的信息，下面是通风系统和光照系统的控制区域，如图 4-1 所示。

图 4-1　系统主界面

知识目标

- 掌握 Spinner 组件、ImageView 组件。
- 掌握 LinearLayout（线性布局）。
- 掌握布局和组件的常见属性。

技能目标

- 能使用线性布局完成布局设计。

4.1 选择主界面布局方式

15 创建系统主界面和线性布局

任务 3 中 Splash 界面的布局使用了 RelativeLayout（相对布局），关于创建布局我们只是了解了点皮毛，各种布局和组件有不同之处，但实际上所有布局和组件也会有很多共同点。

要构建应用让别人使用，一定要保证他们看到的和预想的是一样的。要创建美观的布局，使得界面友好，又易于操作，就需要根据需求来选择适当的布局和 GUI 组件来实现。根据这个页面的特点，可以选择线性布局（LinearLayout），线性布局会在垂直或水平方向上让视图相邻显示。如果是垂直方向，视图会显示在一列上；如果是水平方向，视图会显示在一行上。

4.2 创建线性布局

线性布局的定义可以使用<LinearLayout>元素，如下所示。

```
1. <?xml version="1.0" encoding="utf-8"?>
2. <LinearLayout xmlns:android="http://schemas.android.com/apk/res/android"
3.     xmlns:tools="http://schemas.android.com/tools"
4.     android:layout_width="match_parent"
5.     android:layout_height="match_parent"
6.     android:orientation="vertical"
7.     tools:context=".MainActivity">
8. </LinearLayout>
```

第 4、5 行指定了布局的宽度和高度，这两个属性是必需的，相对布局也使用了这些属性。

第 6 行指定了按垂直方向（vertical）组织视图，也可以根据需要设置为 android:orientation="horizontal"，按水平方向组织视图。一般来讲，这个属性也是必需的。

4.2.1 添加环境监控布局

在定义线性布局时，如果希望组件以什么顺序出现在布局中，就要以什么样的顺序在 XML 中增加视图。所以，如果希望一个文本视图出现在最上面，那么就要先定义这个视图。例如，主界面中显示环境监控信息的文本视图（TextView）。

```
1. <TextView
2.     android:layout_width="match_parent"
3.     android:layout_height="wrap_content"
4.     android:background="@color/colorPrimaryDark"
5.     android:paddingTop="10dp"
6.     android:paddingBottom="10dp"
7.     android:text="@string/environment"
8.     android:textAlignment="center"
9.     android:textColor="@color/colorWhite"
```

```
10.        android:textSize="20sp"/>
```

第 3 行中，android:layout_height="wrap_content"是指定以文本视图内容的大小决定组件的高度。

第 4 行指定了文本视图的背景颜色，colorPrimaryDark 颜色需要在颜色资源文件 colors.xml 中定义，打开 app/src/main/res/values/colors.xml 文件增加该颜色。

```
1. <?xml version="1.0" encoding="utf-8"?>
2. <resources>
3.     <color name="colorPrimaryDark">#848988</color>
4. </resources>
```

在 Android Studio 中，单击行号右边的颜色提示块打开选择颜色窗口，选择想使用的颜色，如图 4-2 所示。

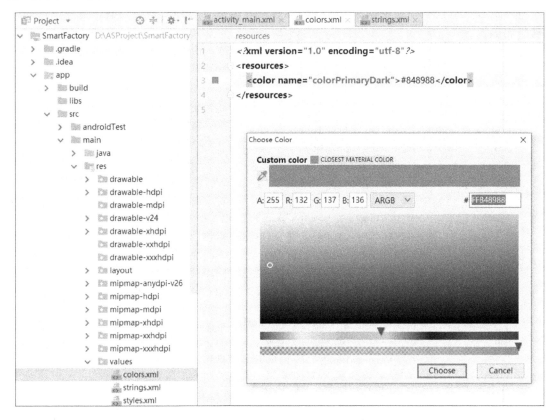

图 4-2 选择颜色

文本视图中第 5 行指定了该文本视图的上内边距为 10dp。通过 paddingTop、paddingBottom、paddingLeft、paddingRight 这些属性可以告诉 Android 视图与其父容器各边之间有多大的间距，如图 4-3 所示。

第 6 行指定了下内边距为 10dp。

第 8 行指定了文本内容居中显示。

第 9 行指定了文字的颜色为 colorWhite，使用这个颜色前要先在 colors.xml 颜色资源文件中定义该颜色。

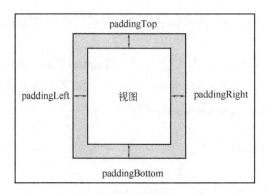

图 4-3　内边距

环境监控文本视图下面要增加一个横向的线性布局放置三种传感器数值。这是一个内嵌的线性布局。

```
1.  <LinearLayout
2.      android:layout_width="match_parent"
3.      android:layout_height="wrap_content"
4.      android:orientation="horizontal"
5.      android:paddingLeft="10dp"
6.      android:paddingTop="20dp"
7.      android:paddingRight="10dp"
8.      android:paddingBottom="20dp">
9.      <TextView
10.         android:layout_width="0dp"
11.         android:layout_height="wrap_content"
12.         android:layout_weight="1"
13.         android:text="@string/temperature_title"
14.         android:textSize="16sp"/>
15.     <TextView
16.         android:id="@+id/tv_temp_value"
17.         android:layout_width="0dp"
18.         android:layout_height="wrap_content"
19.         android:layout_weight="2"
20.         android:text="@string/temperature_value"
21.         android:textColor="@color/colorRed"
22.         android:textSize="16sp"/>
23.     <TextView
24.         android:layout_width="0dp"
25.         android:layout_height="wrap_content"
26.         android:layout_weight="1"
27.         android:text="@string/humility_title"
28.         android:textSize="16sp"/>
29.     <TextView
30.         android:id="@+id/tv_hum_value"
```

```
31.        android:layout_width="0dp"
32.        android:layout_height="wrap_content"
33.        android:layout_weight="2"
34.        android:text="@string/humility_value"
35.        android:textColor="@color/colorRed"
36.        android:textSize="16sp"/>
37.    <TextView
38.        android:layout_width="0dp"
39.        android:layout_height="wrap_content"
40.        android:layout_weight="1"
41.        android:text="@string/light"
42.        android:textSize="16sp"/>
43.    <TextView
44.        android:id="@+id/tv_light_value"
45.        android:layout_width="0dp"
46.        android:layout_height="wrap_content"
47.        android:layout_weight="2"
48.        android:text="@string/light_value"
49.        android:textColor="@color/colorRed"
50.        android:textSize="16sp"/>
51. </LinearLayout>
```

第 4 行指定了线性布局是水平方向的，可以平行地显示视图。这个布局中有六个文本视图，水平方向依次排列。

第 12、19、26、33、40、47 行都使用到了一个属性：

```
android:layout_weight="number"
```

这个属性为视图指定一个权重，为视图分配权重就是告诉它要占多大比重的空间。参数 number 是一个大于 0 的数字。为视图分配权重时要确保视图有足够的空间放置相应的内容。布局为权重值大于或等于 1 的所有视图按比例分配所有额外的空间。

如果读者够仔细的话，可以看到每个文本视图的 android:layout_width 都指定了宽度为 0dp，这样做的原因是什么呢？因为布局分配的是剩余空间，所以一般将组件的宽度设置为 0dp，如果线性布局是垂直方向的话，则将组件的高度设置为 0dp。

在此，权重值总和为 15，第一个组件占比为 1/15，第二个组件占比为 2/15。

4.2.2 添加禁入区域监控布局

下面添加禁入区域监控文本视图，该区域的信息显示使用的也是线性布局，该布局内有一个文本视图。

16 添加禁入区域监控布局

```
1. <TextView
2.     android:layout_width="match_parent"
3.     android:layout_height="wrap_content"
```

```
4.      android:background="@color/colorPrimaryDark"
5.      android:paddingTop="10dp"
6.      android:paddingBottom="10dp"
7.      android:text="@string/breaking_area_monitoring"
8.      android:textAlignment="center"
9.      android:textColor="@color/colorWhite"
10.     android:textSize="20sp"/>
11. <LinearLayout
12.     android:layout_width="match_parent"
13.     android:layout_height="wrap_content"
14.     android:orientation="horizontal"
15.     android:paddingLeft="10dp"
16.     android:paddingTop="20dp"
17.     android:paddingRight="10dp"
18.     android:paddingBottom="20dp">
19.     <TextView
20.         android:id="@+id/tv_break_value"
21.         android:layout_width="match_parent"
22.         android:layout_height="wrap_content"
23.         android:gravity="center"
24.         android:text="@string/breaking_value"
25.         android:textColor="@color/colorRed"
26.         android:textSize="16sp"/>
27. </LinearLayout>
```

注意，不要忘了在字符串资源文件 strings.xml 和颜色资源文件 colors.xml 中添加相应的字符串和颜色。

4.2.3 添加设备控制布局

添加设备控制文本视图，在该视图下面是三个线性布局，每个线性布局中依次放置了一个文本视图、一个 spinner、一个图片视图。

Spinner 组件是 Android 中的值下拉列表。可以利用这个组件从一组值中选择单个值。使用 spinner 组件时，往往要让它显示一个列表值，允许用户选择他们想要的值。到目前为止，我们已经学过使用 strings.xml 文件指定单个字符串值。这里要做的就是指定一个字符串数组，再让 spinner 组件引用这个数组。

要增加一个字符串数组，可以按如下方式创建。

```
1. <string-array name="control_status">
2.     <item>打开</item>
3.     <item>关闭</item>
4.     <item>自动</item>
5. </string-array>
```

这个字符串数组中包含 3 个值，分别是"打开""关闭""自动"，可以根据需要增加更多的值。在布局文件中可以通过以下方式引用。

```
android:entries="@array/control_status"
```

默认会选择 spinner 列表项最上面的值,单击 spinner 查看列表项,单击一个值时,这个值将被选中,如图 4-4 所示。

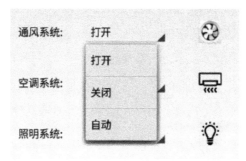

图 4-4 spinner 列表项

图片视图会用到图片文件,我们需要把这些图片文件添加到工程中。可以直接将文件拖到工程的 app/src/main/res/drawable-xhdpi 文件夹中。当为应用增加图片时,需要确定是否要为不同密度的屏幕显示不同的图片。在这里,不论屏幕密度是多少,都使用相同分辨率的图片,所以这里只在一个文件夹中放置一组图片,如图 4-5 所示。如果打算在自己的应用中考虑不同的屏幕密度,可以将适合不同屏幕密度的图片文件放在相应的 drawable*文件夹中。

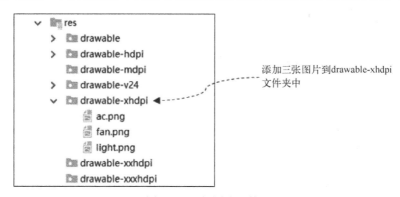

图 4-5 添加图片文件

为工程保存图片时,Android 会为各个图片指定一个 ID,形式为 R.drawable.img_name。例如,将文件 ac.png 的 ID 指定为 R.drawable.ac。

三个线性布局部分的代码如下。

```
1.  <LinearLayout
2.      android:layout_width="match_parent"
3.      android:layout_height="wrap_content"
4.      android:orientation="horizontal"
5.      android:paddingLeft="10dp"
6.      android:paddingTop="20dp"
7.      android:paddingRight="10dp"
8.      android:paddingBottom="20dp">
9.      <TextView
```

```
10.        android:layout_width="0dp"
11.        android:layout_height="wrap_content"
12.        android:layout_weight="1"
13.        android:gravity="center"
14.        android:text="@string/ventilation_system"
15.        android:textSize="16sp"/>
16.    <Spinner
17.        android:layout_width="0dp"
18.        android:layout_height="wrap_content"
19.        android:layout_weight="1"
20.        android:entries="@array/control_status"/>
21.    <ImageView
22.        android:layout_width="0dp"
23.        android:layout_height="wrap_content"
24.        android:layout_weight="1"
25.        android:src="@drawable/fan"/>
26. </LinearLayout>
27. <LinearLayout
28.    android:layout_width="match_parent"
29.    android:layout_height="wrap_content"
30.    android:orientation="horizontal"
31.    android:paddingLeft="10dp"
32.    android:paddingTop="20dp"
33.    android:paddingRight="10dp"
34.    android:paddingBottom="20dp">
35.    <TextView
36.        android:layout_width="0dp"
37.        android:layout_height="wrap_content"
38.        android:layout_weight="1"
39.        android:gravity="center"
40.        android:text="@string/aircondition_system"
41.        android:textSize="16sp"/>
42.    <Spinner
43.        android:layout_width="0dp"
44.        android:layout_height="wrap_content"
45.        android:layout_weight="1"
46.        android:entries="@array/control_status"/>
47.    <ImageView
48.        android:layout_width="0dp"
49.        android:layout_height="wrap_content"
50.        android:layout_weight="1"
51.        android:src="@drawable/ac"/>
52. </LinearLayout>
53. <LinearLayout
54.    android:layout_width="match_parent"
55.    android:layout_height="wrap_content"
```

```
56.     android:orientation="horizontal"
57.     android:paddingLeft="10dp"
58.     android:paddingTop="20dp"
59.     android:paddingRight="10dp"
60.     android:paddingBottom="20dp">
61.     <TextView
62.         android:layout_width="0dp"
63.         android:layout_height="wrap_content"
64.         android:layout_weight="1"
65.         android:gravity="center"
66.         android:text="@string/lighting_system"
67.         android:textSize="16sp"/>
68.     <Spinner
69.         android:layout_width="0dp"
70.         android:layout_height="wrap_content"
71.         android:layout_weight="1"
72.         android:entries="@array/control_status"/>
73.     <ImageView
74.         android:layout_width="0dp"
75.         android:layout_height="wrap_content"
76.         android:layout_weight="1"
77.         android:src="@drawable/light"/>
78. </LinearLayout>
```

第 4 行指定了第一个线性布局为水平方向。

第 12、19、24 行为第一个线性布局的三个组件指定了权重。

第 20、46、72 行指定了 spinner 组件的字符串数组。

到目前为止，完成了主界面的设计。主界面的活动暂时不去实现，主界面中传感器的数据使用的是模拟的数据，执行器的控制也没有实现，这些都等到读者对 Android 的界面设计比较熟练后再来实现。

任务 5　使用活动条导航到全局参数设置界面

任务概述

主界面是应用的中心，应用的其他功能都应该通过主界面快速到达，因此需要设计导航来实现从主界面切换到其他功能界面。本任务介绍使用活动条（ActionBar）来实现导航，为用户提供访问全局参数设置功能的捷径，使应用能够留出更多的空间显示具体内容。

在 Android 中主动导航选项往往会放在活动条中。通常可以在活动顶部看到活动条，常用的动作都可以显示在活动条上。我们可以在主界面和其他活动的顶部增加一个活动条，这个活动条包含一个全局参数设置的按钮，这样不论用户在什么位置都可以轻松访问全局参数设置功能。

知识目标

- 掌握 ActionBar（活动条）概念。
- 掌握 Theme（主题）概念。
- 掌握 AndroidManifest 文件中的常见元素。

技能目标

- 能创建菜单资源文件、样式资源文件。
- 能在活动中将动作项添加到活动条。
- 能在活动中实现动作项单击响应事件。
- 能使用 Intent 传递数据。

5.1　添加活动条和主题

17　添加活动条和主题

活动条有很多用法，可以显示应用或活动名称，使用户知道当前在应用中的哪个位置，可以在活动条上突出显示共享内容或完成搜索，也可以导航到其他活动来完成一个动作。

要增加活动条，需要使用一个包含活动条的主题（Theme）。主题就是应用到整个活动或应用的一个样式，确保应用具有一致性的外观。它会控制多个方面，如活动的背景和活动条的颜色，以及文本的样式等。Android 提供了很多内置主题。

如果希望在 API 11（Android 3.0）或更高级别的版本上运行，可以应用主题 Theme.Holo 或它的子类来增加活动条。对于 API 21（Android 5.0）或更高级别的版本，还可以使用 Theme.Material 主题。可根据你希望的应用外观选择不同的主题，如图 5-1 和图 5-2 所示。

图 5-1　Theme.Holo 主题（API 11 以上）

图 5-2　Material 主题（API 21 以上）

如果选择 API 7 或更低级别的版本，用于支持一些较老的设备，也可增加活动条，但较为复杂。首先要修改活动，让活动继承 android.support.v7.app.ActionBarActivity 类，而不是 android.app.Activity 类，然后必须应用某个 Theme.AppCompat 主题。

```
1. package cn.edu.jsit.smartfactory;
2. import android.support.v7.app.ActionBarActivity;
3.     …
4. public class MainActivity extends ActionBarActivity{
5.     …
6. }
```

我们希望 SmartFactory 应用包含活动条，并且支持至少 API 17 以上版本的设备，所以不需要使用 ActionBarActivity 和 Theme.AppCompat 主题提供向后的兼容性。这里使用一个 Holo 主题，后面还会介绍使用更高级的 Material 主题。

首先需要确保 MainActivity.java 使用 Activity 类而不是 ActionBarActivity，因为使用了 ActionBarActivity 就只能使用 Theme.AppCompat 主题。如下面代码中第 2 行和第 4 行所示。

```
1. package cn.edu.jsit.smartfactory;
2. import android.app.Activity;
3. import android.os.Bundle;
4. public class MainActivity extends Activity {
5.     @Override
6.     protected void onCreate(Bundle savedInstanceState) {
7.         super.onCreate(savedInstanceState);
8.         setContentView(R.layout.activity_main);
9.     }
10. }
```

应用主题可以在 AndroidManifest.xml 中指定。

```
1. <?xml version="1.0" encoding="utf-8"?>
2. <manifest xmlns:android="http://schemas.android.com/apk/res/android"
3.     package="cn.edu.jsit.smartfactory">
4.     <application
5.         android:allowBackup="true"
6.         android:icon="@mipmap/icon_launcher"
7.         android:label="@string/app_name"
8.         android:supportsRtl="true"
9.         android:theme="@style/AppTheme">
10.         <activity
11.             android:name=".MainActivity">
```

```
12.            android:label="@string/main_page"
13.            …
14.        </activity>
15.    </application>
16. </manifest>
```

前面已经提到过，android:theme 属性用于指定主题。在<application>元素中使用这个属性将把主题应用到整个应用。如果在<activity>元素中使用这个属性，则会把它应用到一个活动。

接下来要在样式资源文件中定义样式。创建工程时，Android Studio 会默认创建一个样式资源文件，文件名称为 style.xml，位于 app/src/main/res/values 文件夹中。

```
1. <resources>
2.     <!-- Base application theme. -->
3.     <style name="AppTheme" parent="android:Theme.Holo.Light">
4.         <!-- Customize your theme here. -->
5.     </style>
6. </resources>
```

资源文件包含一个或多个样式。每个样式使用<style>元素定义。每个样式必须有一个名字，使用 name 属性定义。这样 AndroidManifest.xml 中的 android.theme 属性才可以引用这个主题。在这里样式名为 AppTheme，所以 AndroidManifest.xml 中可以使用@style/AppTheme 来引用这个主题。parent 属性指定这个样式从哪里继承，这里我们希望继承 Theme.Holo.Light。

还可以修改一个现有主题的属性，使用样式资源文件定制应用的外观。为此，可以在<style>中增加一个<Item>元素，描述想做的修改。例如

```
<item name="android.background">#FF0000</item>
```

可以将所有活动的背景变成红色。

5.2 创建动作项

18　创建动作项

创建全局参数设置按钮，就是向活动条中增加一个动作项，通过三个步骤完成。

5.2.1 在菜单资源文件中定义动作项

在创建一个包含活动的工程时，Android Studio 会自动创建一个默认的菜单资源文件，文件名称为 menu_main.xml，位于 app/src/main/res/menu 文件夹中，所有菜单资源文件都放在这个文件夹中。如果读者的应用没有包含这个文件夹和文件，请自行创建。

```
1. <menu xmlns:android="http://schemas.android.com/apk/res/android"
2.     xmlns:tools="http://schemas.android.com/tools"
3.     xmlns:app="http://schemas.android.com/apk/res-auto"
4.     tools:context=".MainActivity">
5.     <item android:id="@+id/action_setting"
6.         android:title="@string/setting"
```

```
7.        android:icon="@drawable/setting"
8.        android:orderInCategory="1"
9.        app:showAsAction="never"/>
10. </menu>
```

每个菜单资源文件都有一个根元素<menu>。使用<item>元素为菜单增加菜单项（动作项）。每个动作项使用一个单独的<item>来描述。

第 5 行指定了动作项的 ID，以便在活动代码中引用。

第 6 行指定了动作项的文本。

第 7 行指定了动作项的图标。

第 8 行指定动作项在活动条中出现的顺序。

第 9 行指定是否希望这个动作项出现在活动条中。"never"表示将动作项放到溢出区中，而不要放到活动条中。

应用运行结果如图 5-3 所示。

图 5-3　将动作项放置到溢出区

另外，"ifRoom"表示如果有空间，将这个动作项放在活动条中，如果没有空间就放在溢出区中。"withText"表示包含这个动作项的标题文本。"always"表示总把这个动作项放在主活动条中（可能会导致动作项重叠），如图 5-4 所示。

图 5-4　将动作项放置到活动条

5.2.2　在活动中实现 onCreateOptionsMenu()方法

创建菜单资源文件后，需要实现活动的 onCreateOptionsMenu()方法，将菜单中包含的菜单项添加到活动条。创建活动条的菜单时就会运行这个方法，它有一个参数，这是表示活动条的一个 Menu 对象。

```
1. @Override
2. public boolean onCreateOptionsMenu(Menu menu) {
3.     getMenuInflater().inflate(R.menu.menu_main,menu);
4.     return super.onCreateOptionsMenu(menu);
5. }
```

第 3 行将动作项添加到活动条。menu 是一个 Menu 对象，表示活动条。在 Activity 类中有一个 getMenuInflater() 的函数用来返回这个 Activity 的 MenuInflater 对象，并通过 MenuInflater 对象来设置 menu_main.xml 中定义的 menu 作为该 Activity 的菜单。

第 4 行调用父类 onCreateOptionMenu() 方法。

5.2.3 用 onOptionsItemSelected() 方法响应活动条单击

要在单击活动条中的动作项时让活动做出响应，需要实现 onOptionsItemSelected() 方法。只要单击了活动条中的动作项就会运行这个方法。

```
1. @Override
2. public boolean onOptionsItemSelected(MenuItem menuItem) {
3.     switch(menuItem.getItemId()) {
4.     case R.id.action_setting:
5.         return true;
6.     default:
7.         return super.onOptionsItemSelected(menuItem);
8.     }
9. }
```

onOptionsItemSelected() 方法有一个参数，这是一个 MenuItem 对象，表示活动条上单击的动作项。可以使用 MenuItem 的 getItemId() 方法得到活动条上单击的动作项的 ID，来完成一个适当的动作，如启动一个新的活动。

例如，要创建一个名为 SettingActivity 的新活动，使全局参数设置动作项可以启动这个活动。

首先创建一个新的空活动，指定活动名称为 "SettingActivity"，布局名称为 "activity_setting"，在 AndroidManifest.xml 中修改 SettingActivity 标题为 "全局参数设置"。

每一个活动都必须在 AndroidManifest.xml 中声明。如果一个活动没有在这个文件中声明，系统就不会知道它的存在。如果系统不知道一个活动的存在，这个活动就不会允许运行。要在清单文件中声明一个活动，需要在<application>元素中包含一个<activity>元素。应用中的每个活动都需要有一个相应的<activity>元素。

```
1.  <?xml version="1.0" encoding="utf-8"?>
2.  <manifest xmlns:android="http://schemas.android.com/apk/res/android"
3.      package="cn.edu.jsit.smartfactory">
4.      <application
5.          android:allowBackup="true"
6.          android:icon="@mipmap/icon_launcher"
7.          android:label="@string/app_name"
8.          android:supportsRtl="true"
9.          android:theme="@style/AppTheme">
10.         <activity android:name=".SplashActivity">
11.             <intent-filter>
12.                 <action android:name="android.intent.action.MAIN"/>
13.                 <category android:name="android.intent.category.LAUNCHER"/>
14.             </intent-filter>
```

```
15.        </activity>
16.        <activity
17.            android:name=".MainActivity"
18.            android:label="@string/main_page">
19.        </activity>
20.        <activity
21.            android:name=".SettingActivity"
22.            android:label="@string/globle_params_setting">
23.        </activity>
24.    </application>
25. </manifest>
```

第 10~15 行声明了 SplashActivity。
第 16~19 行声明了 MainActivity。
第 20~23 行声明了 SettingActivity。

android:name 属性这行是必须要有的，用来指定活动的类名。如第 21 行中类名为.SettingActivity，类名前面之所以有一个"."前缀，是因为 Android 会结合类名和包名来得出完全限定类名，这里完全限定类名为 cn.edu.jsit.smartfactory.SettingActivity。android:label 属性为活动指定一个标签，活动运行时，这个标签会在活动条中显示。如果这里不指定标签，Android 会使用应用名。活动声明还包括其他属性，如安全权限、是否允许其他应用中的活动使用等。标签对应的字符串需要添加到字符串资源文件中。

修改 onOptionsItemSelected()方法。

```
1.  @Override
2.  public boolean onOptionsItemSelected(MenuItem menuItem) {
3.      switch(menuItem.getItemId()) {
4.      case R.id.action_setting:
5.          Intent intent=new Intent(MainActivity.this, SettingActivity.class);
6.          startActivity(intent);
7.          return true;
8.      default:
9.          return super.onOptionsItemSelected(menuItem);
10.     }
11. }
```

第 5 行创建了一个意图。
第 6 行启动意图中指定的 SettingActivity 活动。

通过意图告诉 Android 启动活动 SettingActivity，向 Android 传递意图时，Android 会知道以什么顺序或序列启动活动，如果用户在设备上单击 Back（后退）按钮，Android 能准确地知道要退回到哪里。

代码中创建了一个从 MainActivity 跳转到 SettingActivity 的意图，我们还可以向这个意图增加额外的信息，使得指定的目标活动能获取这些信息，并以某种方式做出响应。为此，可以使用 putExtra()方法：

```
intent.putExtra("message",value);
```

这里的 Message 是所传入的值的 String 名，值就是 Value。putExtra()方法是重载的，所以值可以有很多不同的类型，可以是基本类型，如布尔（boolean）或整型（int），也可以是基本类型的数组或是一个字符串（String）。可以反复使用 putExtra()为意图增加多个额外的数据，但一定要确保为每个数据指定一个唯一的名字。

SettingActivity 需要通过某种方法获取 MainActivity 在意图中发送给 Android 的额外信息，可以通过 getIntent()方法实现。getIntent()方法会返回启动活动的意图，可以用这个方法获取随意图发送的额外信息。具体如何做取决于发送的信息的类型。例如：

```
Intent intent=getIntent();//得到意图
```

如果知道意图包含一个 String 值，这个字符串名为"message"，可以使用以下代码：

```
String string=intent.getStringExtra("message");
```

如果意图包含一个 int 值，这个整型数据名为"name"，则可以使用以下代码。其中，default_value 指定了哪个 int 值作为默认值：

```
int intNumber=intent.getIntExtra("name",default_value);
```

更新 MainActivity，将当前温度值增加到意图，传递给 SettingActivity 作为阈值的默认值。

```
1.  @Override
2.  public boolean onOptionsItemSelected(MenuItem menuItem) {
3.      switch(menuItem.getItemId()) {
4.      case R.id.action_setting:
5.          Intent intent=new Intent(MainActivity.this, SettingActivity.class);
6.          TextView tempView=(TextView)findViewById(R.id.tv_temp_value);
7.          String tempValue=tempView.getText().toString();
8.          intent.putExtra("tempValue",tempValue);
9.          startActivity(intent);
10.         return true;
11.     default:
12.         return super.onOptionsItemSelected(menuItem);
13.     }
14. }
```

第 7 行从 ID 为 text_temp_value 的文本视图得到文本（温度值）。

第 8 行将这个文本增加到意图，指定名称为"tempValue"。

要使 SettingActivity 活动启动时获取意图包含的信息并显示。首先要在 activity_setting.xml 文件中增加一个文本视图。

```
1.  <TextView
2.      android:id="@+id/text_temp_value"
3.      android:layout_width="wrap_content"
4.      android:layout_height="wrap_content"
5.      android:text=""/>
```

然后更新 SettingActivity 活动。

```
1. @Override
2. protected void onCreate(Bundle savedInstanceState) {
3.     super.onCreate(savedInstanceState);
4.     setContentView(R.layout.activity_setting);
5.     Intent intent=getIntent();
6.     String tempValue=intent.getStringExtra("tempValue");
7.     TextView tempText=(TextView)findViewById(R.id.text_temp_value);
8.     tempText.setText(tempValue);
9. }
```

第 5 行得到意图。

第 6 行使用 getStringExtra()方法从意图中得到信息。

第 7 行得到显示温度的文本视图。

第 8 行将 getStringExtra()得到的信息增加到文本视图中。

应用运行结果如图 5-5 所示。

图 5-5　跳转到全局参数设置界面

任务 6　创建全局参数设置界面

任务概述

智慧工厂应用不是一个封闭的系统，需要从外界获取数据，或者将应用产生的数据保存到外部系统，这样应用就少不了和外界的系统进行通信，所以一些参数必须事先设置好，才能保证通信的实现。例如，要从物联网云平台获取传感器数据和通过物联网云平台控制通风系统、空调系统、照明系统的打开和关闭，应用就要和物联网云平台直接进行通信，需要设置的参数有物联网云平台的 IP 地址、项目标识、物联网云平台登录的账号和密码，以及各类传感器、执行器 ID。为了实现控制系统的自动控制，需要事先设置好传感器阈值，如温度阈值为 20℃，当温度超过 20℃时空调系统自动打开。另外，应用要通过网络连接智能摄像头，实现工厂作业区的实时监控，就要事先设置好摄像头的 IP 地址。

在任务 5 中已经创建一个名为 SettingActivity 的活动和一个名为 activity_setting 的布局，并实现了从主界面到这个界面的导航，下面要来设计全局参数设置界面的布局，如图 6-1 所示。

知识目标

- 掌握 GridLayout（网格布局）。
- 掌握 EditText 组件、Button 组件。
- 掌握组件单击事件的四种实现方式。
- 掌握自定义组件样式 shape。
- 掌握 Application 类。
- 掌握 SharedPreferences 类。

技能目标

- 能自定义和使用组件的 shape。
- 能使用 SharedPreference 保存参数。
- 能使用用户自定义 Application 对象保存参数。
- 能使用 Device File Explorer（文件管理器）。

6.1　添加网格布局

图 6-1　全局参数设置界面

任务 5 中已经用到了 RelativeLayout（相对布局）、LinearLayout（线性布局），它们各有特点。相对布局会根据相对位置显示布局中的视图，采用这种布局可以相对于布局中的其他视图定义各个视图的位置，也可以相对于其父布局指定视图的位置。线性布局在垂直或水平方向上相邻地显示视图，如果是垂直方向，视图将显示在一列上，如果是水平方向，视图会显示在一行上。

19　添加网格布局 1

网格布局（GridLayout）将屏幕划分为由行、列和单元格组成的网络。可以指定布局分为多少列，视图要放到哪里，以及视图要跨多少行或列。使用网格布局要求使用 API 14 或者更高等级的版本。

要创建新的网络布局，首先要画出这个界面的草图，这样就能清楚地看到需要多少行和多少列，每个视图应该放在哪里，以及各个视图跨几列。现将屏幕划分为 16 列，15 行，然后将各种视图分配到不同的单元格中，如图 6-2 所示。例如，第 0 行上有一个文本视图（内容为"物联网云平台参数设置"），它从第 0 列开始跨 16 列；第 1 行上有一个文本视图（内容为"服务器地址："），它从第 0 列开始跨 6 列，第 1 行上还有一个可编辑的文本域（使用 EditText 组件），它从第 6 列开始跨 11 列；第 13、14 行上有一个按钮，它从第 0 列开始跨 16 列，从第 13 行开始，跨 2 行……

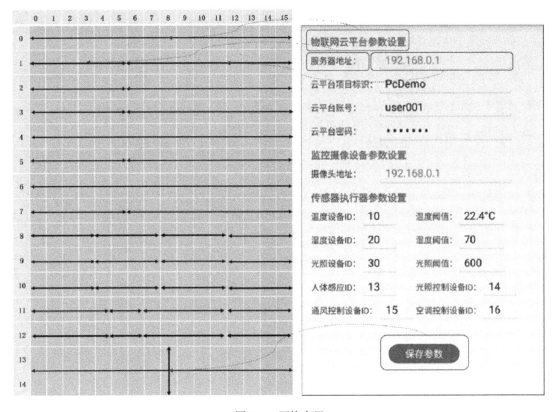

图 6-2 网格布局

定义网格布局和其他类型的布局类似，不过要使用<GridLayout>元素。

```
1. <?xml version="1.0" encoding="utf-8"?>
2. <GridLayout xmlns:android="http://schemas.android.com/apk/res/android"
3.     xmlns:tools="http://schemas.android.com/tools"
4.     android:layout_width="match_parent"
5.     android:layout_height="match_parent"
6.     android:columnCount="16"
7.     android:paddingLeft="16dp"
```

```
8.        android:paddingTop="16dp"
9.        android:paddingRight="16dp"
10.       android:paddingBottom="16dp"
11.       tools:context=".SettingActivity">
12. </GridLayout>
```

第 6 行指定网格布局要使用多少列,这里是 16 列。也可以使用 android:rowCount="number"来指定最大行数,不过实际应用中通常会让 Android 根据布局中的视图个数来确定行数。

由于布局文件较长,这里选取局部进行分析。

```
1. <TextView
2.        android:layout_width="wrap_content"
3.        android:layout_height="wrap_content"
4.        android:layout_row="0"
5.        android:layout_column="0"
6.        android:layout_columnSpan="16"
7.        android:layout_gravity="left"
8.        android:text="cloud_params_setting"
9.        android:textColor="@color/colorBlue"
10.       android:textSize="16sp"
11.       android:textStyle="bold"/>
12. <TextView
13.       android:layout_width="wrap_content"
14.       android:layout_height="wrap_content"
15.       android:layout_row="1"
16.       android:layout_column="0"
17.       android:layout_columnSpan="5"
18.       android:text="@string/server_address"/>
19. <EditText
20.       android:id="@+id/et_server_address"
21.       android:layout_width="wrap_content"
22.       android:layout_height="wrap_content"
23.       android:layout_row="1"
24.       android:layout_column="5"
25.       android:layout_columnSpan="11"
26.       android:ems="11"
27.       android:hint="@string/address_hint"/>
```

第 4 行指定行从 0 开始。

第 5 行指定列从 0 开始。

第 6 行指定该文本视图跨 16 列。

第 7 行指定该文本视图居左。

第 15~17 行指定该文本视图行从 1 开始,列从 0 开始,跨 5 列。

第 23~25 行指定该可编辑文本组件行从 1 开始,列从 5 开始,跨 11 列。

第 26 行将该组件宽度指定为 11 个字符的宽度。

第 27 行指定没有输入内容之前的提示内容。

```
1.  <Button
2.      android:id="@+id/btn_save_params"
3.      android:layout_width="120dp"
4.      android:layout_height="32dp"
5.      android:layout_row="13"
6.      android:layout_column="0"
7.      android:layout_columnSpan="16"
8.      android:layout_rowSpan="2"
9.      android:layout_gravity="center_vertical|center_horizontal"
10.     android:gravity="center_horizontal|center_vertical"
11.     android:background="@drawable/button_style"
12.     android:text="保存参数"
13.     android:textSize="16sp"
14.     android:textColor="@color/colorWhite"/>
```

第 5～8 行指定按钮行从 13 开始，列从 0 开始，跨 16 列，跨 2 行。

第 9 行指定按钮垂直方向居中，并且水平方向居中。

第 10 行指定了按钮文本（"保存参数"）水平方向居中，并且垂直方向居中。

第 11 行指定了按钮背景。这里使用了自定义 shape 圆形按钮。drawable 中可以使用图片，也可以使用自定义的 shape。Android 开发中我们经常要改变组件（如 Button）的背景、颜色、样式等，通常情况下可以直接使用不同的图片来改变组件的样式。但是，如果使用的图片特别多或者比较大，就会导致 APK 文件体积增大，这是非常不友好的。Android 提供了一套自定义图形的方法，也就是 shape。一般用 shape 定义的 XML 文件存放在 drawable 目录下，若项目没有该目录则新建一个，而不要将它放在 drawable-hdpi 等目录中。在 app/src/main/res/drawable 下新建 button_style.xml 文件，如图 6-3 所示。

20　添加网格布局 2

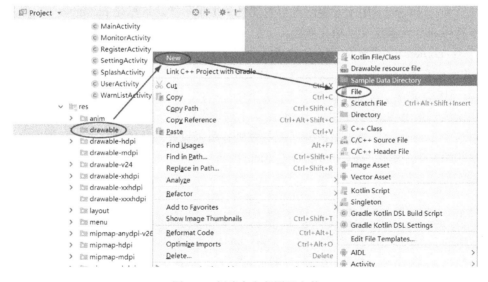

图 6-3　创建自定义图形文件

在新建文件对话框中输入文件名称"button_style.xml"。

```xml
1.  <?xml version="1.0" encoding="utf-8"?>
2.  <selector xmlns:android="http://schemas.android.com/apk/res/android">
3.      <item android:state_pressed="true">
4.          <shape xmlns:android=http://schemas.android.com/apk/res/android android:shape="rectangle">
5.              <!-- 圆角深灰色按钮 -->
6.              <solid android:color="@color/colorPrimaryDark"/>
7.              <corners android:radius="16dip"/>
8.          </shape>
9.      </item>
10.     <item android:state_pressed="false">
11.         <shape xmlns:android="http://schemas.android.com/apk/res/android" android:shape="rectangle">
12.             <!-- 圆角蓝色按钮 -->
13.             <solid android:color="@color/colorBlue"/>
14.             <corners android:radius="16dip"/>
15.         </shape>
16.     </item>
17. </selector>
```

第 2~17 行使用 selector 属性根据不同的选择状态来定义不同的效果。这里为按钮设计了两种效果，当按钮按下去时使用第一种图形显示，当按钮没被按下去时使用第二种图形显示。

第 3~9 行<item>元素定义了按钮按下去的图形。

第 3 行 android:state_pressed 是 selector 的一个属性，表示是否按压，当为 true 时表示按压。另外还有其他属性，如 android:state_selected，表示是否选中，android:state_focused 表示是否获得焦点，android:state_enabled 表示是否响应事件。

第 4 行 android:shape 用于设置组件自身属性的效果形状，这里是 rectangle（矩形）。

第 6 行<solid>元素定义了填充颜色。

第 7 行<corners>元素定义了圆角的半径，值越大，角越圆。单位 dip 指设备独立像素可以使得 UI 元素在不同密度屏幕上看起来一样大小。

第 10~16 行定义了按钮没被按下去的图形。

如果定义了多个 item，要注意 item 是从上往下匹配的，如果匹配到一个 item，那按钮就将采用这个 item，而不是采用最佳匹配的规则。所以设置默认的状态，一定要写在最后，如果写在前面，则后面所有的 item 都不会起作用。

完整布局文件如下。

```xml
1.  <?xml version="1.0" encoding="utf-8"?>
2.  <GridLayout xmlns:android="http://schemas.android.com/apk/res/android"
3.      xmlns:tools="http://schemas.android.com/tools"
4.      android:layout_width="match_parent"
5.      android:layout_height="match_parent"
```

```
6.         android:columnCount="16"
7.         android:paddingLeft="16dp"
8.         android:paddingTop="16dp"
9.         android:paddingRight="16dp"
10.        android:paddingBottom="16dp"
11.        tools:context=".SettingActivity">
12.        <TextView
13.            android:layout_width="wrap_content"
14.            android:layout_height="wrap_content"
15.            android:layout_row="0"
16.            android:layout_column="0"
17.            android:layout_columnSpan="16"
18.            android:layout_gravity="left"
19.            android:text="@string/cloud_params_setting"
20.            android:textColor="@color/colorBlue"
21.            android:textSize="16sp"
22.            android:textStyle="bold"/>
23.        <TextView
24.            android:layout_width="wrap_content"
25.            android:layout_height="wrap_content"
26.            android:layout_row="1"
27.            android:layout_column="0"
28.            android:layout_columnSpan="5"
29.            android:text="@string/server_address"/>
30.        <EditText
31.            android:id="@+id/et_server_address"
32.            android:layout_width="wrap_content"
33.            android:layout_height="wrap_content"
34.            android:layout_row="1"
35.            android:layout_column="5"
36.            android:layout_columnSpan="11"
37.            android:ems="11"
38.            android:hint="@string/address_hint"/>
39.        <TextView
40.            android:layout_width="wrap_content"
41.            android:layout_height="wrap_content"
42.            android:layout_row="2"
43.            android:layout_column="0"
44.            android:layout_columnSpan="5"
45.            android:text="@string/cloud_project_label"/>
46.        <EditText
47.            android:id="@+id/et_project_label"
48.            android:layout_width="wrap_content"
49.            android:layout_height="wrap_content"
50.            android:layout_row="2"
51.            android:layout_column="5"
```

```
52.         android:layout_columnSpan="11"
53.         android:ems="11"
54.         android:text="PcDemo"/>
55.     <TextView
56.         android:layout_width="wrap_content"
57.         android:layout_height="wrap_content"
58.         android:layout_row="3"
59.         android:layout_column="0"
60.         android:layout_columnSpan="5"
61.         android:text="@string/cloud_account"/>
62.     <EditText
63.         android:id="@+id/et_cloud_account"
64.         android:layout_width="wrap_content"
65.         android:layout_height="wrap_content"
66.         android:layout_row="3"
67.         android:layout_column="5"
68.         android:layout_columnSpan="11"
69.         android:ems="11"
70.         android:text="user001"/>
71.     <TextView
72.         android:layout_width="wrap_content"
73.         android:layout_height="wrap_content"
74.         android:layout_row="4"
75.         android:layout_column="0"
76.         android:layout_columnSpan="5"
77.         android:text="@string/cloud_account_password"/>
78.     <EditText
79.         android:id="@+id/et_cloud_account_password"
80.         android:layout_width="wrap_content"
81.         android:layout_height="wrap_content"
82.         android:layout_row="4"
83.         android:layout_column="5"
84.         android:layout_columnSpan="11"
85.         android:ems="11"
86.         android:inputType="textPassword"
87.         android:text="user001"/>
88.     <TextView
89.         android:layout_width="wrap_content"
90.         android:layout_height="wrap_content"
91.         android:layout_row="5"
92.         android:layout_column="0"
93.         android:layout_columnSpan="16"
94.         android:layout_gravity="left"
95.         android:paddingTop="10dp"
96.         android:text="@string/camera_address_setting"
97.         android:textColor="@color/colorBlue"
```

```xml
98.            android:textSize="16sp"
99.            android:textStyle="bold"/>
100.    <TextView
101.            android:layout_width="wrap_content"
102.            android:layout_height="wrap_content"
103.            android:layout_row="6"
104.            android:layout_column="0"
105.            android:layout_columnSpan="6"
106.            android:text="@string/camera_address"/>
107.    <EditText
108.            android:id="@+id/et_camera_address"
109.            android:layout_width="wrap_content"
110.            android:layout_height="wrap_content"
111.            android:layout_row="6"
112.            android:layout_column="6"
113.            android:layout_columnSpan="10"
114.            android:ems="11"
115.            android:hint="@string/address_hint"/>
116.    <TextView
117.            android:layout_width="wrap_content"
118.            android:layout_height="wrap_content"
119.            android:layout_row="7"
120.            android:layout_column="0"
121.            android:layout_columnSpan="16"
122.            android:layout_gravity="left"
123.            android:paddingTop="10dp"
124.            android:text="@string/sensor_actuator_setting"
125.            android:textColor="@color/colorBlue"
126.            android:textSize="16sp"
127.            android:textStyle="bold"/>
128.    <TextView
129.            android:layout_width="wrap_content"
130.            android:layout_height="wrap_content"
131.            android:layout_row="8"
132.            android:layout_column="0"
133.            android:layout_columnSpan="4"
134.            android:text="@string/temp_sensor_id"/>
135.    <EditText
136.            android:id="@+id/et_temp_sensor_id"
137.            android:layout_width="wrap_content"
138.            android:layout_height="wrap_content"
139.            android:layout_row="8"
140.            android:layout_column="4"
141.            android:layout_columnSpan="4"
142.            android:ems="3"
143.            android:text="10"/>
```

```xml
144.    <TextView
145.        android:layout_width="wrap_content"
146.        android:layout_height="wrap_content"
147.        android:layout_row="8"
148.        android:layout_column="8"
149.        android:layout_columnSpan="2"
150.        android:text="@string/temp_threshold_value"/>
151.    <EditText
152.        android:id="@+id/et_temp_threshold_value"
153.        android:layout_width="wrap_content"
154.        android:layout_height="wrap_content"
155.        android:layout_row="8"
156.        android:layout_column="10"
157.        android:layout_columnSpan="6"
158.        android:ems="4"
159.        android:text="30"/>
160.
161.    <TextView
162.        android:layout_width="wrap_content"
163.        android:layout_height="wrap_content"
164.        android:layout_row="9"
165.        android:layout_column="0"
166.        android:layout_columnSpan="4"
167.        android:text="@string/temp_sensor_id"/>
168.    <EditText
169.        android:id="@+id/et_hum_sensor_id"
170.        android:layout_width="wrap_content"
171.        android:layout_height="wrap_content"
172.        android:layout_row="9"
173.        android:layout_column="4"
174.        android:layout_columnSpan="4"
175.        android:ems="3"
176.        android:text="20"/>
177.    <TextView
178.        android:layout_width="wrap_content"
179.        android:layout_height="wrap_content"
180.        android:layout_row="9"
181.        android:layout_column="8"
182.        android:layout_columnSpan="2"
183.        android:text="@string/temp_threshold_value"/>
184.    <EditText
185.        android:id="@+id/et_hum_threshold_value"
186.        android:layout_width="wrap_content"
187.        android:layout_height="wrap_content"
188.        android:layout_row="9"
189.        android:layout_column="10"
```

```xml
190.            android:layout_columnSpan="6"
191.            android:ems="4"
192.            android:text="70"/>
193.        <TextView
194.            android:layout_width="wrap_content"
195.            android:layout_height="wrap_content"
196.            android:layout_row="10"
197.            android:layout_column="0"
198.            android:layout_columnSpan="4"
199.            android:text="@string/light_sensor_id"/>
200.        <EditText
201.            android:id="@+id/et_light_sensor_id"
202.            android:layout_width="wrap_content"
203.            android:layout_height="wrap_content"
204.            android:layout_row="10"
205.            android:layout_column="4"
206.            android:layout_columnSpan="4"
207.            android:ems="3"
208.            android:text="30"/>
209.        <TextView
210.            android:layout_width="wrap_content"
211.            android:layout_height="wrap_content"
212.            android:layout_row="10"
213.            android:layout_column="8"
214.            android:layout_columnSpan="2"
215.            android:text="@string/light_threshold_value"/>
216.        <EditText
217.            android:id="@+id/et_light_threshold_value"
218.            android:layout_width="wrap_content"
219.            android:layout_height="wrap_content"
220.            android:layout_row="10"
221.            android:layout_column="10"
222.            android:layout_columnSpan="6"
223.            android:ems="4"
224.            android:text="1000"/>
225.        <TextView
226.            android:layout_width="wrap_content"
227.            android:layout_height="wrap_content"
228.            android:layout_row="11"
229.            android:layout_column="0"
230.            android:layout_columnSpan="4"
231.            android:text="@string/body_sensor_id"/>
232.        <EditText
233.            android:id="@+id/et_body_sensor_id"
234.            android:layout_width="wrap_content"
235.            android:layout_height="wrap_content"
```

```xml
236.        android:layout_row="11"
237.        android:layout_column="4"
238.        android:layout_columnSpan="4"
239.        android:ems="3"
240.        android:text="13"/>
241.    <TextView
242.        android:layout_width="wrap_content"
243.        android:layout_height="wrap_content"
244.        android:layout_row="11"
245.        android:layout_column="8"
246.        android:layout_columnSpan="4"
247.        android:text="@string/light_controller_id"/>
248.    <EditText
249.        android:id="@+id/et_light_controller_id"
250.        android:layout_width="wrap_content"
251.        android:layout_height="wrap_content"
252.        android:layout_row="11"
253.        android:layout_column="12"
254.        android:layout_columnSpan="4"
255.        android:ems="3"
256.        android:text="14"/>
257.    <TextView
258.        android:layout_width="wrap_content"
259.        android:layout_height="wrap_content"
260.        android:layout_row="12"
261.        android:layout_column="0"
262.        android:layout_columnSpan="5"
263.        android:text="@string/ventilation_controller_id"/>
264.    <EditText
265.        android:id="@+id/et_ventilation_controller_id"
266.        android:layout_width="wrap_content"
267.        android:layout_height="wrap_content"
268.        android:layout_row="12"
269.        android:layout_column="5"
270.        android:layout_columnSpan="2"
271.        android:ems="3"
272.        android:text="15"/>
273.    <TextView
274.        android:layout_width="wrap_content"
275.        android:layout_height="wrap_content"
276.        android:layout_row="12"
277.        android:layout_column="7"
278.        android:layout_columnSpan="5"
279.        android:text="@string/air_controller_id"/>
280.    <EditText
```

```xml
281.        android:id="@+id/et_air_controller_id"
282.        android:layout_width="wrap_content"
283.        android:layout_height="wrap_content"
284.        android:layout_row="12"
285.        android:layout_column="12"
286.        android:layout_columnSpan="4"
287.        android:ems="3"
288.        android:text="16"/>
289.    <Button
290.        android:id="@+id/btn_save_params"
291.        android:layout_width="120dp"
292.        android:layout_height="32dp"
293.        android:layout_row="13"
294.        android:layout_column="0"
295.        android:layout_columnSpan="16"
296.        android:layout_rowSpan="2"
297.        android:layout_gravity="center_vertical|center_horizontal"
298.        android:background="@drawable/button_style"
299.        android:gravity="center_horizontal|center_vertical"
300.        android:text="@string/save_params"
301.        android:textSize="16sp"
302.        android:textColor="@color/colorWhite"
303.        android:onClick="onClickSave"/>
304. </GridLayout>
```

在strings.xml字符串资源文件中增加如下字符串。

```xml
1.  <string name="cloud_params_setting">物联网云平台参数设置</string>
2.  <string name="server_address">服务器地址：</string>
3.  <string name="cloud_project_label">云平台项目标识：</string>
4.  <string name="cloud_account">云平台账号：</string>
5.  <string name="cloud_account_password">云平台密码：</string>
6.  <string name="camera_address_setting">监控摄像设备参数设置</string>
7.  <string name="camera_address">摄像头地址：</string>
8.  <string name="sensor_actuator_setting">传感器执行器参数设置</string>
9.  <string name="temp_threshold_value">温度阈值：</string>
10. <string name="light_sensor_id">光照设备ID：</string>
11. <string name="light_threshold_value">光照阈值：</string>
12. <string name="body_sensor_id">人体感应ID：</string>
13. <string name="light_controller_id">光照控制设备ID：</string>
14. <string name="ventilation_controller_id">通风控制设备ID：</string>
15. <string name="air_controller_id">空调控制设备ID：</string>
16. <string name="save_params">保存参数</string>
17. <string name="temp_sensor_id">温度设备ID：</string>
18. <string name="humi_threshold_value">湿度阈值：</string>
19. <string name="address_hint">192.168.0.1</string>
```

6.2 添加按钮单击事件

在上述布局文件中还需要为按钮添加单击事件。

21 添加按钮单击事件

在布局文件中给按钮添加 onClick 属性来实现单击事件的监控。android:onClick= "onClickSave"，指定该文本视图被单击后要调用的方法名称为 onClickSave()方法，然后在活动中添加 onClickSave()方法，通过这个方法来保存输入的配置信息。这种实现单击事件的方法在任务 3 中已经使用过，这是一种比较容易实现的单击事件处理方法之一，但是这种方法不足的地方是每一次维护的时候都要去布局文件中修改代码。

下面介绍其他的 3 种方法，请大家逐一尝试。

6.2.1 通过匿名内部类实现

第一种方法是通过匿名内部类来实现。在 Android 开发中我们会经常看到各种匿名内部类的使用，那么在实现 button 单击事件的时候也可以用匿名内部类。这样做的好处是不需要重新写一个类，直接在 new 的时候去实现想实现的方法，很方便。另外，当别的地方都用不到这个方法的时候建议使用匿名内部类，匿名内部类的特性之一就是拥有高内聚，高内聚是设计原则之一。其不足之处是当别的地方也需要使用同样的一个方法时还要在那个地方重新写一次匿名内部类，这样使得代码的冗余度很高，也不方便后期的维护。实现方法是，在活动中添加如下代码。

```
1. //绑定 button 按钮
2. btn=(Button)findViewById(R.id.btn_save_params);
3. //监听 button 事件
4. btn.setOnClickListener(new OnClickListener() {
5.     @Override
6.     public void onClick(View v) {
7.         //保存数据
8.     }
9. });
```

6.2.2 通过独立类实现

第二种方法是重新写一个独立的类来实现业务逻辑或是想要的效果。其优点是方便维护，可以降低代码的冗余度，可以同时用到多个地方。其不足之处是当只使用一次时浪费资源，程序的性能不高，当有很多个这类方法时代码的可读性不高，此时不方便维护。实现方法是，首先在活动中添加如下代码。

```
1. private Button btn;
2. @Override
3. protected void onCreate(Bundle savedInstanceState) {
4.     super.onCreate(savedInstanceState);
5.     setContentView(R.layout.activity_main);
6.     btn=(Button)findViewById(R.id.btn_save_params);
```

```
7.      btn.setOnClickListener(new ButtonClick(this));
8.  }
```

然后创建外部类 ButtonClick，ButtonClick 类中需要实现 OnClickListener 接口，并重写其中的方法。

```
1.  public class ButtonClick implements OnClickListener {
2.      private Context context;
3.      //重载 btnClick 方法
4.      public ButtonClick(Context ct) {
5.          this.context=ct;
6.      }
7.      @Override
8.      public void onClick(View v) {
9.        //保存数据
10.     }
11. }
```

6.2.3 通过 OnClickListener 接口实现

第三种方法是通过 OnClickListener 接口实现，并实现其中的 onClick()方法。这种方法与独立类实现的原理是一样的，优点和缺陷也是基本一样的。

```
1.  public class SettingActivity extends Activity implements OnClickListener {
2.      ...
3.      private Button btn;
4.      ...
5.      @Override
6.      protected void onCreate(Bundle savedInstanceState) {
7.          super.onCreate(savedInstanceState);
8.          setContentView(R.layout.activity_setting);
9.          ...
10.         btn=(Button)findViewById(R.id.btn_save_params);
11.         btn.setOnClickListener(this);
12.         ...
13.     }
14.     //实现 OnClickListener 接口中的方法
15.     @Override
16.     public void onClick(View v) {
17.       //保存数据
18.     }
19. }
```

6.3 保存全局参数

下面要考虑如何将这些用户输入的参数保存下来？保存到什么地方？本任务通过两种方

式来保存全局参数，分别是 SharedPreference 和用户自定义 Application 对象。

22 保存全局参数1

6.3.1 使用 SharedPreference 保存参数

Android 提供了一种用来保存这些配置参数的方法，那就是使用 SharedPerference。SharedPreference 是 Android 系统中轻量级存储数据的一种方式，操作简便快捷，它的本质是基于 XML 文件存储 Key-Value 键值对数据，适合存放程序状态的配置信息。如果设置为私有，它会被存储在应用程序的私有存储区，文件权限是私有的，这时它只能供写入者读取。支持的数据类型有 boolean、int、float、long 和 String 等。它存储在应用程序的私有目录（data/data/包名/shared_prefs）下自定义的 XML 文件中。

1．保存数据

使用 SharedPerformance 保存数据的步骤有 4 步。

1）获取应用中的 SharedPreferences 对象。

获取应用中的 SharedPreferences 对象有两种方式。

① 第一种是通过 getSharedPreferences()方法。如果需要多个通过名称参数来区分的 shared preference 文件，名称可以通过第一个参数来指定。

```
SharedPreferences sharedPref=getSharedPreferences("params", Context.MODE_PRIVATE);
```

- ◆ mode 指定为 MODE_PRIVATE，则该配置文件只能被自己的应用程序访问。
- ◆ mode 指定为 MODE_WORLD_READABLE，则该配置文件除了自己访问外还可以被其他应用程序读取。
- ◆ mode 指定为 MODE_WORLD_WRITEABLE，则该配置文件除了自己访问外还可以被其他应用程序读取和写入。

② 第二种是通过 getPreferences()方法。这个方法默认使用当前类不带包名的类名作为文件的名称，同时也和上面一样传入文件读写的模式。

```
SharedPreferences sharedPref=getPreferences(Context.MODE_PRIVATE);
```

2）通过执行 edit()方法创建一个 SharedPreferences.Editor。

```
SharedPreferences.Editor editor=sharedPref.edit();
```

3）通过类似 putString()与 putInt()等方法传递 keys 与 values。

```
editor.putString("server_address",serverAddress.getText().toString().trim());
```

4）通过 commit()方法提交修改。

```
editor.commit();
```

2．读取数据

从 SharedPreferences 文件中读取配置信息，可以通过类似于 getString()及 getInt()等方法来读取。在这些方法里面传递我们想要获取的 value 对应的 key，并提供一个默认的 value 作为查找的 key 不存在时函数的返回值。

```
String serverAddress=sharedPref.getString("server_address", "192.168.0.1");
```

关于 SharedPerformance 其他情况的处理。

1）每个应用有一个默认的偏好文件 preferences.xml，用户可以将配置信息保存到这个文件中，使用 getDefaultSharedPreferences 获取，其余操作是一样的，如下所示。

```
SharedPreferences preferences=PreferenceManager.getDefaultSharedPreferences(this);
```

2）如果 B 应用要读取或写入 A 应用中的 Preference，前提条件是 A 应用中该 preference 创建时指定了 Context.MODE_WORLD_READABLE 或 Context.MODE_WORLD_WRITEABLE 权限，代表其他的应用能读取或写入。首先要在 B 中创建一个指向 A 应用的 Context，如下所示。

```
Context otherAppsContext=createPackageContext("A应用的包名", Context.CONTEXT_IGNORE_SECURITY);
```

然后再通过 context 获取 SharedPreferences 实体，如下所示。

```
SharedPreferences sharedPreferences=otherAppsContext.getSharedPreferences("SharedPreferences的文件名", Context.MODE_WORLD_READABLE);
String name=sharedPreferences.getString("key", "");
```

当然，如果不通过创建 Context 访问其他应用的 preference，也可以以读取 XML 文件的方式直接访问其他应用 preference 对应的 XML 文件，如下所示。

```
File xmlFile=new File("/data/data/<cn.edu.jsit.smartfactory>/shared_prefs/params.xml");
```

3）检索 SharedPerformance 保存的数据是否存在。

```
SharedPreference myPreference=getSharedPreferences("myPreference", Context.MODE_PRIVATE);
boolean isContains=myPreference.contains("key");//检查当前键是否存在
```

在 SettingActivity 活动中创建 initView()方法，初始化视图。在 onClickSave()方法中将用户输入的参数通过 SharedPreference 保存到 params.xml 文件中。

```
1.  private void initView() {
2.      serverAddress=(EditText)findViewById(R.id.et_server_address);
3.      projectLabel=(EditText)findViewById(R.id.et_project_label);
4.      cloudAccount=(EditText)findViewById(R.id.et_cloud_account);
5.      cloudAccountPassword=(EditText)findViewById(R.id.et_cloud_account_password);
6.      cameraAddress=(EditText)findViewById(R.id.et_camera_address);
7.      tempSensorId=(EditText)findViewById(R.id.et_temp_sensor_id);
8.      tempThresholdValue=(EditText)findViewById(R.id.et_temp_threshold_value);
9.      humSensorId=(EditText)findViewById(R.id.et_hum_sensor_id);
```

```
10.     humThresholdValue=(EditText)findViewById(R.id.et_hum_threshold_value);
11.     lightSensorId=(EditText)findViewById(R.id.et_light_sensor_id);
12.     lightThresholdValue=(EditText)findViewById(R.id.et_light_threshold_value);
13.     bodySensorId=(EditText)findViewById(R.id.et_body_sensor_id);
14.     lightControllerId=(EditText)findViewById(R.id.et_light_controller_id);
15.     ventilationControllerId=(EditText)findViewById(R.id.et_ventilation_controller_id);
16.     airControllerId=(EditText)findViewById(R.id.et_air_controller_id);
17.     //初始化温度、湿度、光照值
18.     Intent intent=getIntent();
19.     tempThresholdValue.setText(intent.getStringExtra("tempValue"));
20.     humThresholdValue.setText(intent.getStringExtra("humValue"));
21.     lightThresholdValue.setText(intent.getStringExtra("lightValue"));
22. }
```

第 18～21 行获取从主界面通过意图传递过来的实时温度、湿度、光照值作为阈值的默认值。

```
1.  public void onClickSave(View view) {
2.      SharedPreferences sharedPref=getSharedPreferences( "params", Context.MODE_PRIVATE);
3.      SharedPreferences.Editor editor=sharedPref.edit();
4.      editor.putString("server_address",serverAddress.getText().toString().trim());
5.      editor.putString("project_label",projectLabel.getText().toString().trim());
6.      editor.putString("cloud_account",cloudAccount.getText().toString().trim());
7.      editor.putString("cloud_account_password",cloudAccountPassword.getText().toString().trim());
8.      editor.putString("camera_address",cameraAddress.getText().toString().trim());
9.      editor.putString("temp_sensor_id", tempSensorId.getText().toString().trim());
10.     editor.putString("temp_threshold_value",tempThresholdValue.getText().toString().trim());
11.     editor.putString("hum_sensor_id",humSensorId.getText().toString().trim());
12.     editor.putString("hum_threshold_value", humThresholdValue.getText().toString().trim());
13.     editor.putString("light_sensor_id", lightSensorId.getText().toString().trim());
```

```
    14.         editor.putString("light_threshold_value", lightThresholdValue.getText().toString().trim());
    15.         editor.putString("body_sensor_id", bodySensorId.getText().toString().trim());
    16.         editor.putString("light_controller_id", lightControllerId.getText().toString().trim());
    17.         editor.putString("ventilation_controller_id", ventilationControllerId.getText().toString().trim());
    18.         editor.putString("air_controller_id", airControllerId.getText().toString().trim());
    19.         editor.commit();
    20.         //显示保存成功提示信息
    21.         Toast showToast=Toast.makeText(this,"保存成功! ", Toast.LENGTH_SHORT);
    22.         showToast.setGravity(Gravity.CENTER, 0, 0);
    23.         showToast.show();
    24.    }
```

第 2 行通过前面所讲的第一种方式获取应用中的 SharedPreferences 对象。指定参数文件名称为 param，方式为 MODE_PRIVATE（私有）。

第 3 行创建一个 SharedPreferences.Editor 对象 editor。

第 4~18 行通过 putString()方法将所有参数以 key-value 的方式保存到 param.xml 中。

第 19 行提交保存的参数。

第 21~23 行使用了一个简单的组件 Toast 来显示提示信息"保存成功!"。Toast 是在屏幕上显示的弹出式消息组件，只用来提供信息，用户无法与它进行交互。显示一个 Toast 时，活动会保持可见而且可交互，Toast 到期时会自动消失。

Toast 只能使用活动代码创建，不能在布局中定义。要创建一个 Toast，需要调用 Toast.makeText()方法，并传入 3 个参数，一个 Context（上下文，通常对应当前活动）、一个 CharSequence（要显示的信息）以及一个 int 时间值。一旦创建了 Toast，可以调用它的 show()方法来显示。

第 21 行 this 表示当前活动，显示信息为"保存成功!"，这里使用了内置常量 Toast.LENGTH_SHORT 来指定显示时间（大约 1s），另外一个内置常量是 Toast.LENGTH_LONG（大约 3s）。

第 22 行使用 setGravity()方法来设置 Toast 显示的位置。第一个参数（位置）可以使用 android.view.Gravity 类的常量来指定，有 Gravity.CENTER、Gravity.BOTTOM、Gravity.LEFT、Gravity.RIGHT、Gravity.TOP 等常量。第二个参数是 Toast 位于屏幕 X 轴的位移，大于 0 表示往屏幕右边移动，小于 0 表示往屏幕左边移动。第三个参数与参数二一样，不过是在屏幕 Y 轴的位移，大于 0 表示往屏幕下方移动，小于 0 表示往屏幕上方移动。

应用运行结果如图 6-4 所示。

3．使用 Device File Explorer 查看保存数据

单击"保存参数"按钮后，可以看到"保存成功!"的提示信息。要查看保存到

params.xml 文件中的参数信息，可以使用 Device File Explorer，查看方法如下。

选择"View"→"Tool Windows"→"Device File Explorer"菜单命令，如图 6-5 所示。

图 6-4　保存参数

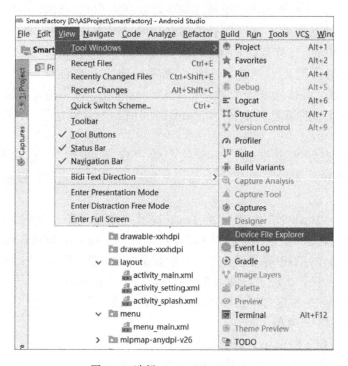

图 6-5　选择 Device File Explorer

打开"Device File Explorer"窗口，在右侧边栏会显示目录浏览窗口，如图 6-6 所示。

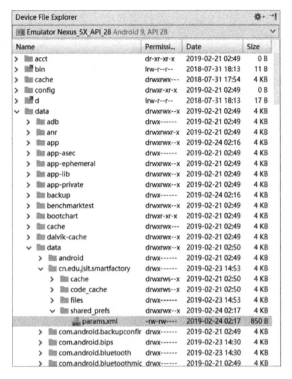

图 6-6　目录浏览

在 data/data/cn.edu.jsit.smartfactory 目录下，可以看到 params.xml 文件，打开该文件可以查看保存的数据，如图 6-7 所示。

图 6-7　param.xml 文件内容

6.3.2 使用用户自定义 Application 保存全局参数

在 Android 中经常使用意图（Intent）来传递数据，但是意图传递的数据类型太少，因此我们经常通过 SharedPreference 来保存和传递全局参数。SharedPreference 也有不足之处，第一是传递的数据类型有限，第二是无法避免多线程访问冲突，且如果使用真机调试的话，SharedPreference 文件不好查看。除了这两种方式以外，还可以通过 Application 来实现多组件之间共享数据。

23 保存全局参数2

Application 是维护全局状态的基类。默认情况下 Android 会为每个应用分配一个进程，进程的名称就是每个项目的包名，当进程启动的时候，系统会创建 Application 对象并调用对象的 onCreate 方法，它诞生于其他任何组件对象之前，并且一直存活，直到应用进程结束。

Application 在项目运行过程中不会改变，而 Activity 在切换过程中，会不断地创建和销毁。

Application 对同一个应用程序是唯一的（静态单例 Singleton Pattern，是 Java 中最简单的设计模式之一），对象全局可访问，且全程陪同应用进程，适合共享全局状态，初始化应用所需的全部服务，因此可以通过 Application 来进行一些数据传递、数据共享和数据缓存等操作。

其用法是，首先编写一个自定义 Application 类继承 Application，重写里面的 onCreate() 方法，其实 android.app.Application 包的 onCreate() 才是真正的 Android 程序的入口点。在包 cn.edu.jsit.smartfactory.tools 下创建 SmartFactoryApplication 类。

```
1.  package cn.edu.jsit.smartfactory.tools;
2.  import android.app.Application;
3.  public class SmartFactoryApplication extends Application {
4.
5.      private String serverAddress="";
6.      private String projectLabel="";
7.      private String cloudAccount="";
8.      private String cloudAccountPassword="";
9.      private String cameraAddress="";
10.     private String tempSensorId="";
11.     private float tempThresholdValue=0;
12.     private String humSensorId="";
13.     private float humThresholdValue=0;
14.     private String lightSensorId="";
15.     private float lightThresholdValue=0;
16.     private String bodySensorId="";
17.     private String lightControllerId="";
18.     private String ventilationControllerId="";
19.     private String airControllerId="";
20.     private boolean isLogin = false;
21.
22.     @Override
23.     public void onCreate() {
24.         super.onCreate();
```

```java
25.      }
26.
27.      public void setServerAddress(String s) {
28.          this.serverAddress=s;
29.      }
30.
31.      public String getServerAddress() {
32.          return this.serverAddress;
33.      }
34.
35.      public void setProjectLabel(String s) {
36.          this.projectLabel=s;
37.      }
38.
39.      public String getProjectLabel() {
40.          return this.projectLabel;
41.      }
42.
43.      public void setCloudAccount(String s) {
44.          this.cloudAccount=s;
45.      }
46.
47.      public String getCloudAccount() {
48.          return this.cloudAccount;
49.      }
50.
51.      public void setCloudAccountPassword(String s) {
52.          this.cloudAccountPassword=s;
53.      }
54.
55.      public String getCloudAccountPassword() {
56.          return this.cloudAccountPassword;
57.      }
58.
59.      public void setCameraAddress(String s) {
60.          this.cameraAddress=s;
61.      }
62.
63.      public String getCameraAddress() {
64.          return this.cameraAddress;
65.      }
66.
67.      public void setTempSensorId(String s) {
68.          this.tempSensorId=s;
69.      }
70.
```

```java
71.     public String getTempSensorId() {
72.         return this.tempSensorId;
73.     }
74.
75.     public void setTempThresholdValue(float s) {
76.         this.tempThresholdValue=s;
77.     }
78.
79.     public float getTempThresholdValue() {
80.         return this.tempThresholdValue;
81.     }
82.
83.     public void setHumSensorId(String s) {
84.         this.humSensorId=s;
85.     }
86.
87.     public String getHumSensorId() {
88.         return this.humSensorId;
89.     }
90.
91.     public void setHumThresholdValue(float s) {
91.         this.humThresholdValue=s;
93.     }
94.
95.     public float getHumThresholdValue() {
96.         return this.humThresholdValue;
97.     }
98.
99.     public void setLightSensorId(String s) {
100.        this.lightSensorId=s;
101.    }
102.
103.    public String getLightSensorId() {
104.        return this.lightSensorId;
105.    }
106.
107.    public void setLightThresholdValue(float s) {
108.        this.lightThresholdValue=s;
109.    }
110.
111.    public float getLightThresholdValue() {
112.        return this.humThresholdValue;
113.    }
114.
115.    public void setBodySensorId(String s) {
116.        this.bodySensorId=s;
```

```
117.        }
118.
119.        public String getBodySensorId() {
120.            return this.bodySensorId;
121.        }
122.
123.        public void setLightControllerId(String s) {
124.            this.lightControllerId=s;
125.        }
126.
127.        public String getLightControllerId() {
128.            return this.lightControllerId;
129.        }
130.
131.        public void setVentilationControllerId(String s) {
132.            this.ventilationControllerId=s;
133.        }
134.
135.        public String getVentilationControllerId() {
136.            return this.ventilationControllerId;
137.        }
138.
139.        public void setAirControllerId(String s) {
140.            this.airControllerId=s;
141.        }
142.
143.        public String getAirControllerId() {
144.            return this.airControllerId;
145.        }
146.
147.        public void setIsLogin(boolean s) {
148.            this.isLogin=s;
149.        }
150.
151.        public boolean getIsLogin() {
152.            return this.isLogin;
153.        }
154.    }
```

第 5~20 行定义了静态变量。

第 22~25 行重写了 Application 类的 onCreate()方法。

第 24 行调用了父类的 onCreate()方法，这是必需的，此外在 onCreate()内可以加入需要的代码。

对每个变量定义了 setXXX()用于设置变量值，getXXX()方法用于获取变量值，实现了封装。

接下来，修改活动 SettingActivity，在活动中定义一个 SmartFactoryApplication 对象，在单击事件中先将用户输入的参数值保存的该对象并检查关键参数是否为空，然后将 SmartFactoryApplication 对象中的参数通过 SharedPreference 保存到 params.xml 文件中。

修改 SettingActivity 类，代码如下：

```
1.  package cn.edu.jsit.smartfactory;
2.  import android.app.Activity;
3.  import android.content.Context;
4.  import android.content.Intent;
5.  import android.content.SharedPreferences;
6.  import android.os.Bundle;
7.  import android.view.View;
8.  import android.widget.EditText;
9.  import android.widget.Toast;
10. import android.view.Gravity;
11. import java.util.Timer;
12. import java.util.TimerTask;
13. import cn.edu.jsit.smartfactory.tools.CloudHelper;
14. import cn.edu.jsit.smartfactory.tools.SmartFactoryApplication;
15.
16. public class SettingActivity extends Activity {
17.     private EditText serverAddress;
18.     private EditText projectLabel;
19.     private EditText cloudAccount;
20.     private EditText cloudAccountPassword;
21.     private EditText cameraAddress;
22.     private EditText tempSensorId;
23.     private EditText tempThresholdValue;
24.     private EditText humSensorId;
25.     private EditText humThresholdValue;
26.     private EditText lightSensorId;
27.     private EditText lightThresholdValue;
28.     private EditText bodySensorId;
29.     private EditText lightControllerId;
30.     private EditText ventilationControllerId;
31.     private EditText airControllerId;
32.     private SmartFactoryApplication smartFactory;
33.
34.     @Override
35.     protected void onCreate(Bundle savedInstanceState) {
36.         super.onCreate(savedInstanceState);
37.         setContentView(R.layout.activity_setting);
38.         smartFactory=(SmartFactoryApplication) getApplication();
39.         initView();
40.     }
41.
```

```java
42.     private void initView() {
43.         serverAddress=(EditText)findViewById(R.id.et_server_address);
44.         projectLabel=(EditText)findViewById(R.id.et_project_label);
45.         cloudAccount=(EditText)findViewById(R.id.et_cloud_account);
46.         cloudAccountPassword=(EditText)findViewById(R.id.et_cloud_account_password);
47.         cameraAddress=(EditText)findViewById(R.id.et_camera_address);
48.         tempSensorId=(EditText)findViewById(R.id.et_temp_sensor_id);
49.         tempThresholdValue=(EditText)findViewById(R.id.et_temp_threshold_value);
50.         humSensorId=(EditText)findViewById(R.id.et_hum_sensor_id);
51.         humThresholdValue=(EditText)findViewById(R.id.et_hum_threshold_value);
52.         lightSensorId=(EditText)findViewById(R.id.et_light_sensor_id);
53.         lightThresholdValue=(EditText)findViewById(R.id.et_light_threshold_value);
54.         bodySensorId=(EditText)findViewById(R.id.et_body_sensor_id);
55.         lightControllerId=(EditText)findViewById(R.id.et_light_controller_id);
56.         ventilationControllerId=(EditText)findViewById(R.id.et_ventilation_controller_id);
57.         airControllerId=(EditText)findViewById(R.id.et_air_controller_id);
58.         //初始化温度、湿度、光照值
59.         Intent intent=getIntent();
60.         Float tempValue=intent.getFloatExtra("tempValue", 0);
61.         Float humValue=intent.getFloatExtra("humValue", 0);
62.         Float lightValue=intent.getFloatExtra("lightValue", 0);
63.         tempThresholdValue.setText(tempValue.toString());
64.         humThresholdValue.setText(humValue.toString());
65.         lightThresholdValue.setText(lightValue.toString());
66.     }
67.
68.     public void onClickSave(View view) {
69.         smartFactory.setServerAddress(serverAddress.getText().toString().trim());
70.         smartFactory.setProjectLabel(projectLabel.getText().toString().trim());
71.         smartFactory.setCloudAccount(cloudAccount.getText().toString().trim());
72.         smartFactory.setCloudAccountPassword(cloudAccountPassword.getText().toString().trim());
73.         smartFactory.setCameraAddress(cameraAddress.getText().toString().trim());
74.         smartFactory.setTempSensorId(tempSensorId.getText().toString().trim());
```

```
75.            smartFactory.setTempThresholdValue(Float.parseFloat(tempThresholdValue.getText().toString().trim()));
76.            smartFactory.setHumSensorId(humSensorId.getText().toString().trim());
77.            smartFactory.setHumThresholdValue(Float.parseFloat(humThresholdValue.getText().toString().trim()));
78.            smartFactory.setLightSensorId(lightSensorId.getText().toString().trim());
79.            smartFactory.setLightThresholdValue(Float.parseFloat(lightThresholdValue.getText().toString().trim()));
80.            smartFactory.setBodySensorId(bodySensorId.getText().toString().trim());
81.            smartFactory.setLightControllerId(lightControllerId.getText().toString().trim());
82.            smartFactory.setVentilationControllerId(ventilationControllerId.getText().toString().trim());
83.            smartFactory.setAirControllerId(airControllerId.getText().toString().trim());
84.        if(!checkInput(smartFactory)) {
85.            return;
86.        } else {
87.            SharedPreferences sharedPref=getSharedPreferences("params", Context.MODE_PRIVATE);
88.            SharedPreferences.Editor editor=sharedPref.edit();
89.            editor.putString("server_address", smartFactory.getServerAddress());
90.            editor.putString("project_label", smartFactory.getProjectLabel());
91.            editor.putString("cloud_account", smartFactory.getCloudAccount());
92.            editor.putString("cloud_account_password", smartFactory.getCloudAccountPassword());
93.            editor.putString("camera_address", smartFactory.getCameraAddress());
94.            editor.putString("temp_sensor_id", smartFactory.getTempSensorId());
95.            editor.putFloat("temp_threshold_value", smartFactory.getTempThresholdValue());
96.            editor.putString("hum_sensor_id", smartFactory.getHumSensorId());
97.            editor.putFloat("hum_threshold_value", smartFactory.getTempThresholdValue());
98.            editor.putString("light_sensor_id", smartFactory.getLightSensorId());
99.            editor.putFloat("light_threshold_value", smartFactory.getLightThresholdValue());
```

```
100.                    editor.putString("body_sensor_id", smartFactory.getBodySensor
Id());
101.                    editor.putString("light_controller_id", smartFactory.getLight-
ControllerId());
102.                    editor.putString("ventilation_controller_id", smartFactory.get
VentilationControllerId());
103.                    editor.putString("air_controller_id", smartFactory.getAirCo-
ntrollerId());
104.                    editor.commit();
105.                    //显示保存成功提示信息
106.                    showToast(R.string.save_params_sucess);
107.                    Intent intent=new Intent(this, MainActivity.class);
108.                    startActivity(intent);
109.                }
110.            }
111.
112.        private boolean checkInput(SmartFactoryApplication smartFactory) {
113.            boolean result=true;
114.            if(smartFactory.getServerAddress().equals("")) {
115.                showToast(R.string.server_address_empty);
116.                return false;
117.            }
118.            if(smartFactory.getProjectLabel().equals("")) {
119.                showToast(R.string.cloud_project_empty);
120.                return false;
121.            }
122.            if(smartFactory.getCloudAccount().equals("")) {
123.                showToast(R.string.cloud_account_empty);
124.                return false;
125.            }
126.            if(smartFactory.getCloudAccountPassword().equals("")) {
127.                showToast(R.string.cloud_account_password_empty);
128.                return false;
129.            }
130.            if(smartFactory.getCameraAddress().equals("")) {
131.                showToast(R.string.camera_address_empty);
132.                return false;
133.            }
134.            return result;
135.        }
136.
137.        private void showToast(int resId) {
138.            Toast showToast;
139.            showToast=Toast.makeText(this, resId, Toast.LENGTH_SHORT);
140.            showToast.setGravity(Gravity.CENTER, 0, 0);
141.            showToast.show();
```

```
142.        }
143.    }
```

使用 CheckInput()方法完成对用户输入参数的检查，其中服务器地址、云平台项目标识、云平台账号、云平台密码、摄像头地址是必须输入的。

ShowToast()方法用来显示提示信息，帮助用户正确输入参数。保存成功后返回到主界面。

在字符串资源文件中添加如下字符串。

```
1. <string name="server_address_empty">请输入服务器地址！</string>
2. <string name="cloud_project_empty">请输入云平台项目标识！</string>
3. <string name="cloud_account_empty">请输入云平台账号！</string>
4. <string name="cloud_account_password_empty">请输入云平台密码！</string>
5. <string name="camera_address_empty">请输入摄像头地址！</string>
6. <string name="save_params_success">保存成功！</string>
```

下面还需要在 AndroidManifest.xml 文件中增加 android:name 属性指定自定义 Application。

```
1. <application
2.     ...
3.     android:name=".tools.SmartFactoryApplication"
4.     ...
5.     >
6. </application>
```

更新 MainActivity 中的 onOptionsItemSelected()方法。

```
1. @Override
2. public boolean onOptionsItemSelected(MenuItem menuItem){
3.     switch (menuItem.getItemId()){
4.         case R.id.action_setting:
5.             Intent intent = new Intent(MainActivity.this, SettingActivity.class);
6.             TextView tempView = (TextView)findViewById(R.id.tv_temp_value);
7.             float tempValue = Float.parseFloat(tempView.getText().toString().trim());
8.             TextView humView = (TextView)findViewById(R.id.tv_humility_value);
9.             float humValue =Float.parseFloat(humView.getText().toString().trim());
10.            TextView lightView = (TextView)findViewById(R.id.tv_light_value);
11.            float lightValue = Float.parseFloat(lightView.getText().toString().trim());
12.            intent.putExtra("tempValue",tempValue);
```

```
13.            intent.putExtra("humValue",humValue);
14.            intent.putExtra("lightValue",lightValue);
15.            startActivity(intent);
16.            return true;
17.       default:
18.            return super.onOptionsItemSelected(menuItem);
19.     }
20. }
```

任务 7　从云平台获取传感器数据并显示

任务概述

在主界面环境监控区域中如何显示传感器的实时数据？在任务 1 的系统概述中我们知道，传感器采集到数据后，通过无线传感网络传到物联网网关，然后通过网关将数据传到云平台，移动端通过网络访问云平台获取传感器的实时数据并显示，如图 7-1 和图 7-2 所示。本任务就来实现这个功能。

图 7-1　主界面

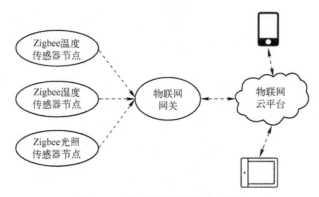

图 7-2　传感器数据采集拓扑图

知识目标
- 掌握 UI 线程、子线程概念。
- 掌握 Handler 传递消息机制。
- 掌握定时器。
- 掌握第三方 jar 包的使用。
- 了解 Android 回调函数机制。
- 了解物联网云平台。

技能目标
- 能导入和使用第三方 jar 包。
- 能从物联网云平台获取传感器数据。

24　添加 jar 包和网络访问权限

7.1 使用第三方提供的 jar 包

访问云平台可以通过第三方提供的 jar 包来实现，因此要将 jar 包引入到工程项目。引入 jar 包的方式有两种。

第一种方式是先切换到 project 目录，将 nlecloudII.jar 包直接复制到 app/libs 目录下，如图 7-3 所示。

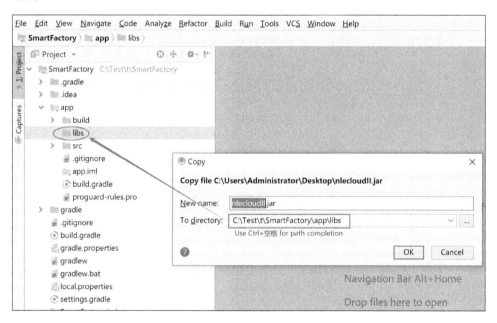

图 7-3　复制 jar 包到 libs 目录下

然后右击该 jar 包，在弹出的快捷菜单中选择 "Add As Library" 命令，如图 7-4 所示。

打开 app 目录下的 build.gradle 文件，可以看到 jar 包已经添加到依赖中，如图 7-5 所示。

第二种方式是在 Android Studio 的 "File" 菜单下，选择 "Project Structure" 命令，为应用引入库。在 "Dependencies" 选项卡中加入需要的 jar 包，如图 7-6 所示。

图 7-4　添加依赖

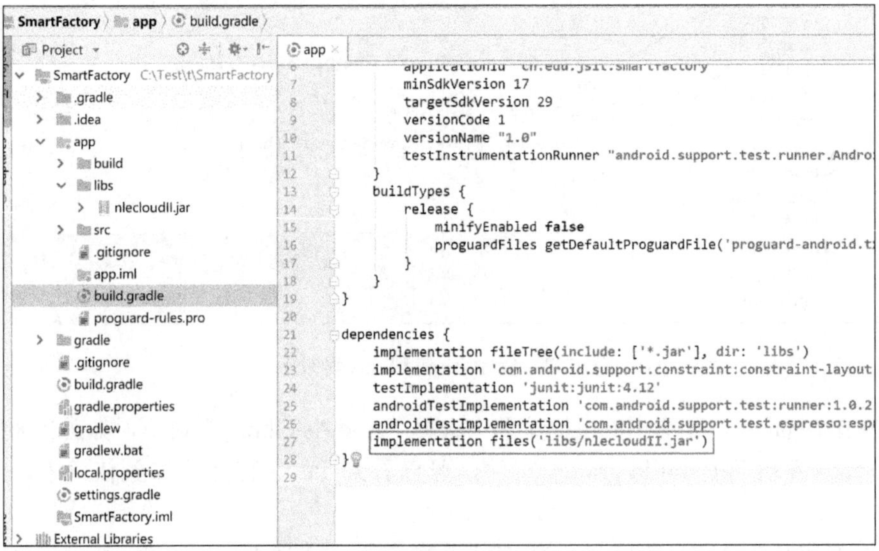

图 7-5　build.gradle 文件内容

按同样的操作，加入另外一个 jar 包 goson-2.2.4.jar，如图 7-7 所示。

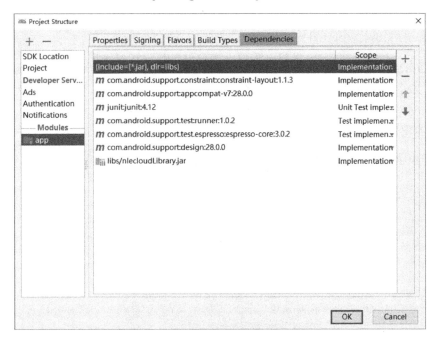

图 7-6 在 Project Structure 中添加 jar 包

图 7-7 引入 gson-2.2.4.jar 包

7.2 添加网络权限

由于需要网络通信，所以要在应用中加入网络权限。Android 访问网络的权限是 android.permission.INTERNET。网络权限在 AndroidManifest.xml 文件中声明。在 Application 元素之前增加如下<user-permission>元素，为应用赋予网络访问权限。

```
<uses-permission android:name="android.permission.INTERNET"/>
```

Android 9.0 默认使用 https 访问，因此如果服务器支持 https 访问，则直接将 http 改成

103

https 即可，如果服务器不支持 https，需要在 application 元素内添加 useClearTextTraffic 属性，并设为 true。

```
android:usesCleartextTraffic="true"
```

如果 Android Studio 模拟器无法联网，首先关闭虚拟机，然后进入 SDK 文件夹下的 emulator 文件夹，按住<Shift>键，选择"在此处打开 Powershell 窗口"命令，如图 7-8 所示。

在打开的窗口中输入如下命令，列出虚拟机名称，结果如图 7-9 所示。

```
.\emulator.exe -list-avds
```

图 7-8 打开 PowerShell 窗口

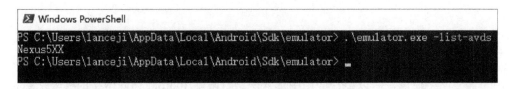

图 7-9 列出虚拟机名称

更改虚拟机 DNS，使其与开发计算机 DNS 一致，如 192.168.0.1。输入如下命令，结果如图 7-10 所示。

```
.\emulator.exe -avd Nexus5XX –dns-server 192.168.0.1
```

图 7-10 更改虚拟机 DNS

7.3 创建 CloudHelper 帮助类

25 创建 CloudHelper 帮助类

创建 CloudHelper 类用于访问物联网云平台。

```java
1.  package cn.edu.jsit.smartfactory.tools;
2.
3.  import android.content.Context;
4.  import android.widget.Toast;
5.  import cn.com.newland.nle_sdk.requestEntity.SignIn;
6.  import cn.com.newland.nle_sdk.responseEntity.SensorInfo;
7.  import cn.com.newland.nle_sdk.responseEntity.User;
8.  import cn.com.newland.nle_sdk.responseEntity.base.BaseResponseEntity;
9.  import cn.com.newland.nle_sdk.util.NCallBack;
10. import cn.com.newland.nle_sdk.util.NetWorkBusiness;
11.
12. public class CloudHelper {
13.
14.     private String token = "";
15.
16.     public String getToken() {
17.         return this.token;
18.     }
19.
20.     public void signIn(final Context c, String address, String account, String pwd) {
21.         NetWorkBusiness nb = new NetWorkBusiness("", address);
22.         nb.signIn(new SignIn(account, pwd),
23.                 new NCallBack<BaseResponseEntity<User>>(c) {
24.                     @Override
25.                     protected void onResponse(BaseResponseEntity<User> response) {
26.                         if (response.getStatus() == 0) {
27.                             token = response.getResultObj().getAccessToken();
28.                         } else {
29.                             Toast.makeText(c, response.getMsg(), Toast.LENGTH_SHORT).show();
30.                         }
31.                     }
32.                 });
33.     }
```

```
34.
35.     public interface DCallback {
36.         void trans(String s);
37.     }
38.
39.     public void getSensorData(Context c, String address, String prjLabel,
40.                         String sensorId, final  DCallback dCallback) {
41.         NetWorkBusiness nb = new NetWorkBusiness(token, address);
42.         nb.getSensor(prjLabel, sensorId,
43.             new NCallBack<BaseResponseEntity<SensorInfo>>(c) {
44.             @Override
45.             protected void onResponse(BaseResponseEntity<SensorInfo> arg0) {
46.                 if (arg0 != null && arg0.getResultObj() != null
47.                         && arg0.getResultObj().getValue() != null)
48.                     dCallback.trans(arg0.getResultObj().getValue());
49.             }
50.         });
51.     }
52.
53. }
```

在 CloudHelper 中定义了 signIn()、getSensorDate()、getToken()三个方法和一个接口 DCallback。

signIn()是登录云平台的方法，其中 c 是指调用该方法的活动对象，address 为物联网云平台的地址，account 是云平台登录账号，pwd 为账号密码。

本书使用新大陆提供的物联网云平台，具体部署过程可以访问中国大学 MOOC 在线开放课程（移动应用开发）。本任务中物联网云平台地址 http://api.nlecloud.com，物联网云平台账号需要到 www.nlecloud.com 进行注册。

第 21 行创建了一个 NetWorkBusiness 对象。

第 22、23 行中调用了 NetWorkBussiness 中的 signIn()方法，该方法中第一个参数为 SignIn()对象，第二个参数为实现了 NCallBack 抽象类的匿名类对象，这里 NcallBack 是一个抽象类。Android 中使用匿名内部类可以不用特意去写一个类去实现这个接口的方法，直接在实例化的时候就写好这个方法（接口、抽象类不能实例化，所以采用匿名内部类的方式来写）。匿名内部类对象可以作为实际参数进行传递，使用一次之后会自动被垃圾回收机制销毁，节省了手机的内存，所以在 Android 中使用普遍。

第 24～31 行重写了回调方法 onResponse()，如果用户正常登录则返回一个 token，它服务端生成的一串字符串，以作客户端进行请求的一个令牌，当第一次登录后，服务器生成一个 token 便将此 token 返回给客户端，以后客户端只须带上这个 token 前来请求数据即可，无须再次带上用户名和密码。

getSensorData()方法是从云平台获取传感器数据的方法。其中，c 是指调用该方法的活动对象；address 为物联网云平台的地址；projectLabel 为项目标识（在新大陆云平台 2.0 中为设备 ID，例如"86423"，如图 7-11 所示）；sensorId 为传感器设备 ID（在新大陆云平台 2.0 中为传感器标识名，例如"z_light"，如图 7-11 所示）；dCallback 是为传递获取到的传感器数据的 DCallback 接口引用。

图 7-11　设备 ID 和传感器标识名

第 35~37 行定义了 DCallback 接口。

第 41 行通过登录后获取到的 token 登录。

第 42、43 行中调用了 NetWorkBussiness 中的 getSensor()方法，该方法中第一个参数为项目标识，第二个参数为传感器设备 ID，第三个参数为实现了 NCallBack 抽象类的匿名类对象。

第 44~49 行重写了回调方法 onResponse()，如果正常返回传感器数据，则通过 DCallback 接口中定义的回调方法 trans()返回该数据，MainActivity 中会重写该方法。这里使用接口回调来传递数据的原因是，通过调用 getSensor()方法从云平台获取数据会有延时。

7.4　从云平台获取传感器数据并在主界面更新

在 SettingActivity 中将全局参数存放到了 SmartFactoryApplication 对象中，在 MainActivity 中从该对象中获取访问物联网云平台的参数，使用 CloudHelper 中的方法登录云平台并获取传感器数据，在主界面上实时更新，如图 7-12 所示。

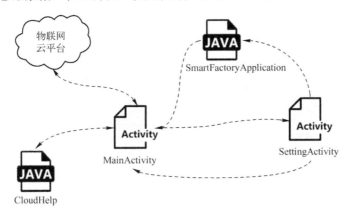

图 7-12　传感器数据获取流程

7.4.1 通过 Handler 机制实现线程消息传递

26 获取传感器数据并更新1

在 Android 中主线程是不可以进行一些耗时的操作的。因为当主线程超过 5 秒无响应时，程序就会弹出程序无响应（Application Not Responding，ANR）对话框。但是在日常开发中，又避免不了去做一些耗时的操作、如访问网络、文件操作等。主线程因为上述原因，却不能轻松完成这些操作，那怎么办？这时候就需要开启新的子线程。在子线程中获取传感器数据后，需要将数据在主界面上更新，但非 UI 线程（主线程）是不能更新 UI 的，那么，应该怎么办呢？不要担心，Android 提供了一种 Handler 机制（消息传递的机制），来帮助我们将子线程的数据传递给主线程，如图 7-13 所示。

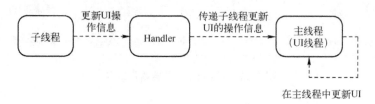

图 7-13 Handler 消息传递机制

在详细了解这个机制前，需要了解以下概念，如表 7-1 所示。

表 7-1 Handle 相关概念

概念	定义	作用
主线程（UI 线程）	当应用第一次启动时，会同时开启一个主线程	处理 UI 相关的事件，如更新、操作等
子线程（工作线程）	手动开启的线程	执行耗时操作，如网络访问、数据加载等
消息（Message）	线程间通信的数据单元，即 Handler 接收、处理的消息对象	存储需操作的通信信息
消息队列（Message Queue）	一种数据结构，以先进先出方式存储数据	存储 Handler 发过来的消息
处理者（Handler）	主线程和子线程之间的通信媒介，线程消息的主要处理者	添加消息到消息队列，处理循环器分配过来的消息
循环器（Looper）	消息队列（Message Queue）与处理者（Handler）之间的通信媒介	消息循环，包括消息出队（循环取出消息队列的消息）和消息分发（将取出的消息发送给对应的处理者）

Handler 机制的工作流程主要包含 4 个步骤，如图 7-14 所示。

1．异步通信准备

在主线程中创建 Looper 对象、Message Queue 对象、Handler 对象，Looper、Message Queue 都属于主线程。创建 Message Queue 后，Looper 进入自动循环。Handler 自动绑定主线程的 Looper 和 Message Queue。

2．消息入队

子线程通过 Handler 发送消息到消息队列，消息内容为子线程对 UI 线程的操作。

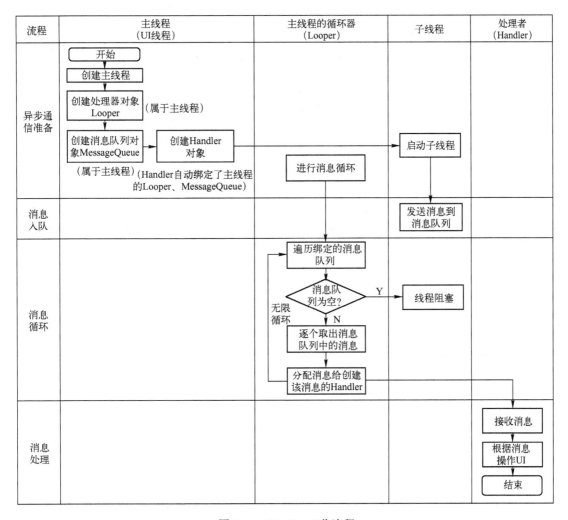

图 7-14 Handler 工作流程

3．消息循环

消息循环包括消息出队（循环取出消息队列的消息）和消息分发（将取出的消息发送给对应的处理者），在消息循环过程中，如果消息队列为空，则线程阻塞。每个线程中只能有一个 Looper，一个 Looper 可以绑定多个线程的 Handler，多个线程可向一个 Looper 所持有的 Message Queue 中发送消息，提供了线程间通信的可能。

4．消息处理

处理者接收处理循环器发送过来的消息，同时处理者根据消息进行 UI 操作。

7.4.2 使用定时器定时更新主界面数据

更新 MainActivity 活动如下。

27 获取传感器数据并更新 2

```
1. public class MainActivity extends Activity {
2.     ...
```

```java
3.      private Spinner spVentilation;
4.      CloudHelper cloudHelper;
5.      SmartFactoryApplication smartFactory;
6.
7.      final Handler handler = new Handler() {
8.          public void handleMessage(android.os.Message msg) {
9.              switch (msg.what) {
10.                 case 1:
11.                     tvLightValue.setText(lightValue + "lx");
12.                     tvTempValue.setText(tempValue + "°C");
13.                     tvHumValue.setText(humValue + "%RH");
14.                 default:
15.                     break;
16.             }
17.         }
18.     };
19.
20.     @Override
21.     protected void onCreate(final Bundle savedInstanceState) {
22.         ...
23.         loadCloudData();
24.     }
25.
26.     private void loadCloudData() {
27.         smartFactory = (SmartFactoryApplication) getApplication();
28.         cloudHelper = new CloudHelper();
29.         if (smartFactory != null &&
30.             smartFactory.getServerAddress() != "" &&
31.             smartFactory.getCloudAccount() != "" &&
32.             smartFactory.getCloudAccountPassword() != "") {
33.           cloudHelper.signIn(getApplicationContext(),
34.                 smartFactory.getServerAddress(),
35.                 smartFactory.getCloudAccount(),
36.                 smartFactory.getCloudAccountPassword());
37.         }
38.         new Timer().schedule(new TimerTask() {
39.             @Override
40.             public void run() {
41.                 if (cloudHelper.getToken() != "") {
42.                     cloudHelper.getSensorData(getApplicationContext(),
43.                         smartFactory.getServerAddress(),
44.                         smartFactory.getProjectLabel(),
45.                         smartFactory.getLightSensorId(),
46.                         new CloudHelper.DCallback() {
47.                             @Override
48.                             public void trans(String s) {
49.                                 lightValue = s;
```

```
50.                             Log.d("lightValue", s);
51.                         }
52.                     });
53.             cloudHelper.getSensorData(getApplicationContext(),
54.                     smartFactory.getServerAddress(),
55.                     smartFactory.getProjectLabel(),
56.                     smartFactory.getTempSensorId(),
57.                     new CloudHelper.DCallback() {
58.                         @Override
59.                         public void trans(String s) {
60.                             tempValue = s;
61.                             Log.d("tempValue", s);
62.                         }
63.                     });
64.             cloudHelper.getSensorData(getApplicationContext(),
65.                     smartFactory.getServerAddress(),
66.                     smartFactory.getProjectLabel(),
67.                     smartFactory.getHumSensorId(),
68.                     new CloudHelper.DCallback() {
69.                         @Override
70.                         public void trans(String s) {
71.                             humValue = s;
72.                             Log.d("humValue", s);
73.                         }
74.                     });
75.             handler.sendEmptyMessage(1);
76.         }
77.     }
78. }, 0, 5000);
79. }
80. ...
81. }
```

第 8～17 行以匿名内部类方式重写了 handleMessage()方法，创建了 Handler 消息处理器对象。可以看到 handleMessage()方法传入了一个参数 Message，该参数就是从子线程传递过来的信息。通过 Message 的 what 属性值来判断是否更新传感器数据。

第 23 行在 onCreate()方法中调用了 loadCloudData()。

第 26～79 行定义了 loadCloudData()方法。

第 27 行通过 getApplication()方法获得当前 Application 对象。

第 28 行创建 CloudHelper 对象。

第 29～37 行从 SmartFactory Application 对象中获取物联网云平台的地址、账号、密码，调用 CloudHelper 类中的 signIn()方法登录云平台。

第 38～78 行中使用了 Timer 和 TimerTask 构建了一个定时任务（子线程）。Timer 是一种线程设施，用于安排以后在后台线程中执行的任务，可安排任务执行一次或者定期重复执行，可以看成一个定时器，可以调度 TimerTask。TimerTask 是一个抽象类，实现了 Runnable

接口，所以具备多线程的能力。

第 38 行使用 Timer 的 schedule 方法制定任务计划。schedule(TimerTasktask,long delay, long period)，该方法中第一个参数是 TimerTask 对象，这里通过匿名内部类方式重写了 run() 方法。第二个参数是延时时间，单位为毫秒（ms），该延时指用户调用 schedule()方法后要等待多长的时间才执行 run()方法，如果该参数为 0，则表示没有延时。第三个参数是任务执行周期，单位是毫秒（ms）。在 run()方法中，如果登录云平台成功，则通过 getSensorValue()方法每隔 5s 从云平台获取温度、湿度、光照度传感器数据。getSensorValue()方法的第三个参数为实现了 DCallBack 抽象类的匿名类对象，在这个匿名类中重写了回调方法 trans()，getSensorValue()方法通过该回调方法传回传感器数据，如图 7-15 所示。

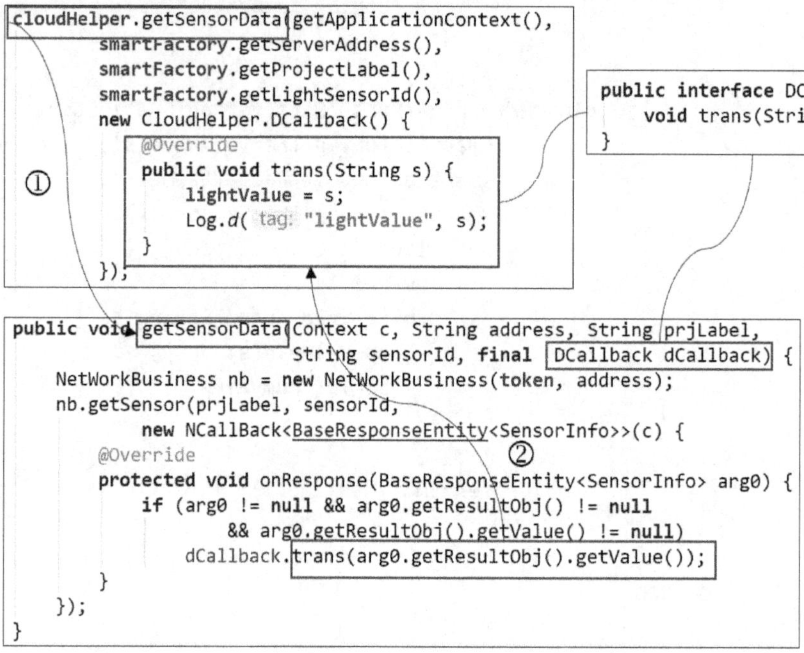

```
① cloudHelper.getSensorData(getApplicationContext(),
        smartFactory.getServerAddress(),
        smartFactory.getProjectLabel(),
        smartFactory.getLightSensorId(),
        new CloudHelper.DCallback() {
            @Override
            public void trans(String s) {
                lightValue = s;
                Log.d( tag: "lightValue", s);
            }
        });

public interface DCallback {
    void trans(String s);
}

public void getSensorData(Context c, String address, String prjLabel,
                String sensorId, final DCallback dCallback) {
    NetWorkBusiness nb = new NetWorkBusiness(token, address);
    nb.getSensor(prjLabel, sensorId,
            new NCallBack<BaseResponseEntity<SensorInfo>>(c) {   ②
                @Override
                protected void onResponse(BaseResponseEntity<SensorInfo> arg0) {
                    if (arg0 != null && arg0.getResultObj() != null
                            && arg0.getResultObj().getValue() != null)
                        dCallback.trans(arg0.getResultObj().getValue());
                }
            });
}
```

图 7-15　接口回调传递传感器数据

第 75 行使用消息处理器 Handler 的 sendEmptyMessage(int what)方法发送消息给主线程。该方法只发送一个带有 what 属性的消息（message），Message 用一个 what 标志来区分不同消息的身份，这样不同的 Handler 使用相同的消息不会弄混，一般使用十六进制形式来表示。该消息用于在 handlerMessage()方法中判断是否更新传感器数据。

这里 sendEmptyMessage()方法的源代码如下。

```
1. public final boolean sendEmptyMessage(int what) {
2.     return sendEmptyMessageDelayed(what, 0);
3. }
```

可以看到该方法调用了 sendEmptyMessageDelay()方法，只是延时为 0。sendEmptyMessageDelayed()方法的源代码如下。

```
1. public final boolean sendEmptyMessageDelayed(int what, long delayMillis) {
```

```
2.      Message msg=Message.obtain();
3.      msg.what=what;
4.      return sendMessageDelayed(msg, delayMillis);
5. }
```

Message.obtain()从 Message 对象池中获取 Message 对象，然后赋值 what 属性值，接着调用 sendMessageDelayed(msg, delayMillis)，此时 delayMillis 是 0ms。

Message 常用的几个属性如下。

◇ Message.what，用来标识信息的 int 值，通过该值主线程能判断出来自不同地方的信息来源。
◇ Message.arg1、Message.arg2，Message 初始定义的用来传递 int 类型值的两个变量。
◇ Message.obj，用来传递任何实例化对象。

sendMessageDelayed()方法的源代码如下。

```
1. public final boolean sendMessageDelayed(Message msg, long delayMillis) {
2.      if(delayMillis<0) {
3.          delayMillis=0;
4.      }
5.      return sendMessageAtTime(msg, SystemClock.uptimeMillis()+delayMillis);
6. }
```

SystemClock.uptimeMillis()是获取系统从开机启动到现在的时间，期间不包括休眠的时间，这里得到的时间是一个相对的时间，而不是通过获取当前的时间（绝对时间）。

sendMessageAtTime()方法的源代码如下。

```
1. public boolean sendMessageAtTime(Message msg, long uptimeMillis) {
2.      boolean sent=false;
3.      MessageQueue queue=mQueue;
4.      if(queue!=null) {
5.          msg.target=this;
6.          sent=queue.enqueueMessage(msg, uptimeMillis);
7.      } else {
8.          RuntimeException e=new RuntimeException(
9.              this+"sendMessageAtTime() called with no mQueue");
10.         Log.w("Looper", e.getMessage(), e);
11.     }
12.     return sent;
13. }
```

所有的消息都是在这里进行入队操作的，当消息队列 MessageQueue 不为空的时候，指定消息对象是本身，然后入队，入队成功后返回布尔值 sent，如果消息成功地放置在消息队列中的话，sent 就返回 true，如果失败则返回 false，其中 enqueueMessage()方法在接收到一个消息之后，根据发送消息的时间来进行队内的轮转计算。

以上结合例子简单分析了 Android 源代码的调用过程。从上面的分析可以看出，发送空消息在底层上的实现并不是没有消息体，它还是会从消息池中获取消息对象，赋值 what 属性的。

任务 8　通过云平台控制执行器

任务概述

执行器在硬件连接上与传感器方式不同，但对于应用来说是透明的，开发者只和物联网云平台进行通信来控制执行器。虽然如此，开发者也应该尽量多地了解系统的硬件拓扑结构，以便更好地了解用户需求，更好地构建应用。

各控制设备和继电器相连，ADAM-4150 通过 485 总线连到物联网网关，Zigbee 节点（含继电器模块）通过 Zigbee 网络连到物联网网关，物联网云平台从移动终端接收控制指令并将指令发送给网关实现对各执行器的控制，如图 8-1 所示。

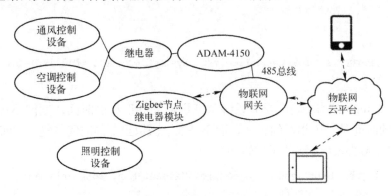

图 8-1　系统控制拓扑图

知识目标
- 掌握 Adapter（适配器）。
- 掌握资源文件的使用。
- 掌握 setResult 和 onActivityResult 机制。

技能目标
- 能通过物联网云平台控制执行器。

28　创建执行器控制方法并设置状态

8.1　创建执行器控制方法

CloudHelper 类中已经包含了 signIn()、getSensorData()方法，实现登录云平台、从云平台获取传感器数据的方法。下面还要在 CloudHelper 类中添加 onOff()方法用于打开和关闭执行器。

```
1.  public void onOff(final Context c,String address,String prjLabel,
2.                    String controllerId,int state){
3.      NetWorkBusiness nb = new NetWorkBusiness(token, address);
```

```
4.     nb.control(prjLabel, controllerId, state,
5.             new NCallBack<BaseResponseEntity>(c) {
6.         @Override
7.         protected void onResponse(BaseResponseEntity response) {
8.             Toast.makeText(c, "控制器操作成功",Toast.LENGTH_SHORT).show();
9.         }
10.    });
11. }
```

第 1、2 行中 onOff()方法的第一个参数是 c 是指调用该方法的活动对象。第二个参数 address 是为物联网云平台的地址。第三个参数 prjLabel 是项目标识（设备 ID）。第四个参数 controllerId 是执行器 ID（执行器标识名）。第五个参数 state 是执行器的状态，如果 state 为 1，则打开执行器；如果 state 为 0，则关闭执行器。

第 3 行创建了一个 NetWorkBusiness 对象。

第 4~10 行中调用了 NetWorkBussiness 中的 control()方法，该方法的第一个参数是项目标识，第二个参数是执行器 ID，第三个参数是执行器状态，第四个参数为实现了 NCallBack 抽象类的匿名类对象。

第 6~9 行重写了回调方法 onResponse()，如果执行器操作成功，则用 Toast 显示"控制器操作成功"。

8.2 使用适配器设置执行器控制状态

在 MainActivity 中用户通过 Spinner 选择"打开""关闭""自动"选项来控制执行器的状态。在任务 4 中我们使用 android:entries="@array/control_status"来引用列表项。之所以可以这么做，是因为数据保存为一个静态字符串数组资源，这个数组在字符串数组资源文件（strings.xml）中描述，只有当数据是 strings.xml 中的一个静态数组时才能使用 android:entries，如果不是这样一个静态数组就无法使用。

如果要将 Spinner 与 strings.xml 以外的其他数据源中的数据绑定，需要采用另外一种不同的方法，要编写活动代码来绑定数据。在这里我们先将 3 种控制状态对应的字符串放到字符串数组 controlStatus 中。controlStatus 数组就是一个数据源，这时这个字符串数组要在 Spinner 列表项中显示就需要使用适配器（Adapter）。适配器就相当于数据源与 Spinner 之间的一座桥，如图 8-2 所示。

图 8-2 适配器（Adapter）

适配器在数据源和 Spinner 之间搭建了一个桥梁。适配器运行 Spinner 显示各种数据源的数据（本书用 controlStatus 数组作为数据源，不过也可以使用数据库或 Web 服务）。

首先创建 controlStatus 数组。

```
1. Resources res=getResources();
2. String[] controlStatus=res.getStringArray(R.array.control_status);
```

在 Android 应用程序中,除了应用程序的编码,还需要关注各种各样的资源,诸如用到的各种静态内容、位图、颜色、布局定义、用户界面字符串、动画等。这些资源一般放置在项目的 res/目录下的独立子目录中。res/目录下的各种子目录中包含了所有的资源,如表 8-1 所示。

表 8-1 Android 资源

目录	资源类型
drawable/	图片文件,如.png、.jpg、.gif 或者 XML 文件,被编译为位图、状态列表、形状、动画图片。它们被保存在 res/drawable/文件夹下,通过 R.drawable 类访问
layout/	定义用户界面布局的 XML 文件。它们被保存在 res/layout/ 文件夹下,通过 R.layout 类访问
menu/	定义应用程序菜单的 XML 文件,如选项菜单、上下文菜单、子菜单等。它们被保存在 res/menu/文件夹下,通过 R.menu 类访问
mipmap/	应用程序的图标
raw/	任意的文件以它们的原始形式保存。需要根据名为 R.raw.filename 的资源 ID,通过调用 Resource.openRawResource()来打开 raw 文件
values/	包含简单值(如字符串、整数、颜色值等)的 XML 文件。这里有一些文件夹下的资源命名规范。1) arrays.xml 代表数组资源,通过 R.array 类访问;2) integers.xml 代表整数资源,通过 R.integer 类访问;3) bools.xml 代表布尔值资源,通过 R.bool 类访问;4) colors.xml 代表颜色资源,通过 R.color 类访问;5) dimens.xml 代表维度值,通过 R.dimen 类访问;6) strings.xml 代表字符串资源,通过 R.string 类访问;7) styles.xml 代表样式资源,通过 R.style 类访问

Android 应用程序被编译后,自动生成一个 R 文件类,其中包含了所有 res/目录下资源的 ID,可以使用 R 类,通过子类+资源名或者直接使用资源 ID 来访问资源。

上面的代码中,第 1 行通过 getResources()方法创建一个 Resources 对象。

第 2 行从 strings.xml 中获取 control_status,创建字符串数组 controlStatus。

使用数组适配器时,首先要初始化这个数组适配器,再把它关联到 Spinner。要初始化数组适配器,首先指定与 Spinner 绑定的数组中包含什么类型的数据。然后传入 3 个参数,分别是 Context(通常是当前活动)、布局资源(指定如何显示数组中的各个项)以及数组本身。

```
1. spVentilation=(Spinner)findViewById(R.id.sp_ventilation_control);
2. final ArrayAdapter<String> adapter=new ArrayAdapter<String>(this,
android.R.layout.simple_list_item_1,controlStatus);
3. spVentilation.setAdapter(adapter);
```

第 1 行获取 Spinner 对象 spVentilation。

第 2 行完成数组适配器的初始化。指定 Spinner 绑定的数组中包含 String 类型数据,第一个参数是 this;第二个参数 simple_list_item_1 是系统内置的布局资源,它告诉适配器在一个文本视图中显示数组的各个项;第三个参数是字符串数组 controlStatus。

第 3 行使用 setAdapter()方法将这个数组适配器与 Spinner 关联。在底层,数组适配器取数组的各个项,用 toString()方法将它转为一个字符串,再把各个结果分别放在一个文本视图中，然后把各个文本视图显示为一个 Spinner 中的一行。

下面要实现一个事件监听器，让 Spinner 中的列表项响应单击事件。事件监听器允许监听应用中发生的事件，如单击视图、视图得到或失去焦点，或者用户按下设备上的一个按键。通过实现事件监听器，可以知道用户什么时候完成某个特定的动作（如选择 Spinner 中的一项），并做出响应。

如果希望 Spinner 中的列表项响应单击事件，需要创建一个 OnItemSelected Listener，并实现它的 onItemSelected()方法。OnItemSelectedListener 会监听什么时候单击了列表项，可以在 onItemSelected()方法中指出活动如何响应这个单击事件。onItemSelected()方法有很多参数，可以使用这些参数找出单击了哪一项，如所单击的视图项的一个引用，它在 Spinner 中的位置（从 0 开始），以及底层数据的行 ID。

```
1.   spVentilation.setOnItemSelectedListener(new AdapterView.OnItemSelectedListener() {
2.       @Override
3.       public void onItemSelected(AdapterView<?> parent,
4.                                  View view,
5.                                  int position,
6.                                  long id) {
7.           Context c = getApplicationContext();
8.           String address = smartFactory.getServerAddress();
9.           String projLabel = smartFactory.getProjectLabel();
10.          String controllerId = smartFactory.getVentilationControllerId();
11.          String status = spVentilation.getItemAtPosition(position).toString();
12.          if (cloudHelper.getToken() != "") {
13.              switch (status) {
14.                  case "打开":
15.                      cloudHelper.onOff(c, address, projLabel, controllerId, 1);
16.                      break;
17.                  case "关闭":
18.                      cloudHelper.onOff(c, address, projLabel, controllerId, 0);
19.                      break;
20.                  case "自动":
21.                      if (Float.parseFloat(tempValue) > smartFactory.getTempThresholdValue()) {
22.                          cloudHelper.onOff(c, address, projLabel, controllerId, 1);
23.                      } else {
24.                          cloudHelper.onOff(c, address, projLabel, controllerId, 0);
25.                      }
26.                      break;
27.                  default:
28.                      cloudHelper.onOff(c, address, projLabel, controllerId, 0);
29.                      break;
30.              }
31.          }
32.      }
```

```
33.
34.      @Override
35.      public void onNothingSelected(AdapterView<?> parent) {
36.      }
37.  });
38.  spVentilation.setSelection(1, true);
```

第 1 行使用 setOnItemSelectedListener()方法实现将创建的 onItemSelectedListener 关联到 Spinner spVentilation。OnItemSelectedListener 是一个内嵌在 AdapterView 类中的类。Spinner 是 AdapterView 的一个子类。

第 3~6 行 onItemSelected()方法中的第一个参数是单击动作发生的视图（这里就是 Spinner），第二个参数是被单击的视图项，第三个参数是被单击视图项在 Spinner 中的位置，第四个参数是被单击视图项的行 ID。

第 7~10 行获得当前的应用程序的全局 Context、云平台服务器地址、项目标识、执行器 ID。

第 11 行根据被单击视图的位置通过 getItemAtPosition()方法获取该被单击视图项的内容，并通过 toString()方法转成字符串赋给 status。

第 12 行通过根据 token 是否为空判断当前是否已经登录云平台。

第 13~30 行根据 status 的不同情况来控制执行器。当 status 为"打开"时，使用 CloudHelper 类中的 onOff()方法打开通风控制设备。当 status 为"关闭"时，关闭通风控制设备。当 status 为"自动"时，将当前温度和温度阈值作比较，如果当前温度大于温度阈值，则打开通风控制设备，否则关闭通风控制设备。

第 34~36 行中重写了 onNothingSelected()方法，该方法在 adapter 为空的时候被调用，此时关闭通风控制设备。

第 38 行使用 Spinner 的 setSelection()方法设置默认情况下选择为"关闭"。第一个参数是指 Spinner 视图项的行 ID，第二个参数为动画设置，仅当请求的位置已经在屏幕上的某个位置时有效。

另外两个 Spinner——spAc 和 spLight 的代码类似。

```
1.  spAc = (Spinner) findViewById(R.id.sp_air_control);
2.  spAc.setAdapter(adapter);
3.  spAc.setOnItemSelectedListener(new AdapterView.OnItemSelectedListener() {
4.      @Override
5.      public void onItemSelected(AdapterView<?> parent,
6.                                 View view,
7.                                 int position,
8.                                 long id) {
9.          Context c = getApplicationContext();
10.         String address = smartFactory.getServerAddress();
11.         String projLabel = smartFactory.getProjectLabel();
12.         String controllerId = smartFactory. getAirControllerId();
13.         String status = spAc.getItemAtPosition(position).toString();
14.         if (cloudHelper.getToken() != "") {
```

```
15.            switch (status) {
16.                case "打开":
17.                    cloudHelper.onOff(c, address, projLabel, controllerId, 1);
18.                    break;
19.                case "关闭":
20.                    cloudHelper.onOff(c, address, projLabel, controllerId, 0);
21.                    break;
22.                case "自动":
23.                    if (Float.parseFloat(humValue) > smartFactory.getTempThresholdValue()) {
24.                        cloudHelper.onOff(c, address, projLabel, controllerId, 1);
25.                    } else {
26.                        cloudHelper.onOff(c, address, projLabel, controllerId, 0);
27.                    }
28.                    break;
29.                default:
30.                    cloudHelper.onOff(c, address, projLabel, controllerId, 0);
31.                    break;
32.            }
33.        }
34.    }
35.
36.    @Override
37.    public void onNothingSelected(AdapterView<?> parent) {
38.    }
39. });
40. spAc.setSelection(1, true);
41.
42. spLight = (Spinner) findViewById(R.id.sp_light_control);
43. spLight.setAdapter(adapter);
44. spLight.setOnItemSelectedListener(new AdapterView.OnItemSelectedListener() {
45.    @Override
46.    public void onItemSelected(AdapterView<?> parent,
47.                               View view,
48.                               int position,
49.                               long id) {
50.        Context c = getApplicationContext();
51.        String address = smartFactory.getServerAddress();
52.        String projLabel = smartFactory.getProjectLabel();
53.        String controllerId = smartFactory.getLightControllerId();
54.        String status = spLight.getItemAtPosition(position).toString();
55.        if (cloudHelper.getToken() != "") {
56.            switch (status) {
57.                case "打开":
58.                    cloudHelper.onOff(c, address, projLabel, controllerId, 1);
59.                    break;
```

```
60.                  case "关闭":
61.                      cloudHelper.onOff(c, address, projLabel, controllerId, 0);
62.                      break;
63.                  case "自动":
64.                      if (Float.parseFloat(lightValue) < smartFactory.getTemp
ThresholdValue()) {
65.                          cloudHelper.onOff(c, address, projLabel, controllerId, 1);
66.                      } else {
67.                          cloudHelper.onOff(c, address, projLabel, controllerId, 0);
68.                      }
69.                      break;
70.                  default:
71.                      cloudHelper.onOff(c, address, projLabel, controllerId, 0);
72.                      break;
73.              }
74.          }
75.      }
76.
77.      @Override
78.      public void onNothingSelected(AdapterView<?> parent) {
79.      }
80. });
81. spLight.setSelection(1, true);
```

8.3 使用 setResult 和 onActivityResult 机制实现返回

任务 6 中在保存好全局参数后，使用意图（Intent）返回到主界面。

29　实现返回

```
1. Intent intent=new Intent(this, MainActivity.class);
2. startActivity(intent);
```

使用意图返回 Android 会重新创建 MainActivity，控制器的状态重新被初始化为默认状态。为了使返回后控制器的状态不变（如原来是"打开"，返回后还是"打开"），则不能使用意图返回主界面，可以使用 setResult 和 onActivityResult 机制。

使用该机制首先需要更新 MainActivity。

```
1.  public class MainActivity extends Activity {
2.      ...
3.      static final private int GET_CODE=0;
4.      ...
5.      @Override
6.      public boolean onOptionsItemSelected(MenuItem menuItem) {
7.          switch(menuItem.getItemId()) {
8.              case R.id.action_setting:
9.                  ...
10.                 Intent intent=new Intent(MainActivity.this,SettingActivity.class);
```

```
11.          Bundle bundle=new Bundle();
12.          bundle.putString("uid",MainActivity.this.toString());
13.          intent.putExtras(bundle);
14.          startActivityForResult(intent, GET_CODE);
15.          return true;
16.     default:
17.          return super.onOptionsItemSelected(menuItem);
18.     }
19. }
20.
21. @Override
22. protected void onActivityResult(int requestCode, int resultCode, Intent data) {
23.     if(requestCode==GET_CODE) {
24.         if(resultCode==RESULT_OK) {
25.             loadCloudData();
26.         }
27.     }
28. }
29. }
```

第 3 行定义了一个整型常量 GET_CODE=0。

第 10 行定义了一个意图,从主界面跳转到全局参数设置界面。

第 11~13 行创建了一个 Bundle 对象,用于保存 MainActivity 活动的标识字符串,其格式为:Activity 的 name+@+当前 object 的索引。

第 14 行使用 startActivityForResult()方法来实现跳转,第一个参数为 intent 对象,第二个参数 requestCode 在目标 Activity 结束时会返回给 onActivityResult()方法,在 onActivityReuslt()方法中可以通过判断 requestCode 的值来确定从哪个目标 Activity 返回。

第 21~28 行重写了 onActivityResult()方法。该方法在目标 Activity 结束返回时被调用,第一个参数 requestCode 用以判断从哪个目标 Activity 返回,第二个参数 resultCode 由目标 Activity 通过 setResult()方法返回,第三个参数 intent 对象可以带有返回的数据。

第 23 行通过返回的 requestCode 与 GET_CODE 对比,确认是否从 SettingActivity 返回。

第 24 行判断 resultCode 是否为 RESULT_OK,resultCode 在 SettingActivity 中通过 setResult()设置。

第 25 行调用 loadCloudData()方法从云平台获取数据并更新主界面。

在 SettingActivity 中还需要更新 onCreate()方法和 onClickSave()方法。

```
1. @Override
2. protected void onCreate(Bundle savedInstanceState) {
3.     ...
4.     Bundle bundle=getIntent().getExtras();
5.     if(bundle!=null) {
6.         uid=bundle.getString("uid");
7.     }
```

```
8.      ...
9.  }
10.
11. public void onClickSave(View view) {
12.     ...
13.     setResult(RESULT_OK,(new Intent()).setAction(uid));
14.     finish();
15. }
```

第 4~7 行获取从 MainActivity 传递过来的活动标识字符串 uid。

第 13 行通过 setResult()方法设置返回结果。第一个参数为结果返回码，值为 RESULT_OK；第二个参数为 intent 对象，通过 setAction()方法设置返回活动标识。

任务 9　创建执行器状态动画

任务概述

在主界面中当我们打开或关闭执行器时，希望右侧的图片能够以动画的形式来显示。例如，通风控制设备打开后，风扇能够转动，空调控制设备打开后，可以模拟送风动画，这样可以让用户对打开这个动作有更加直观的感觉。

知识目标
- 掌握 Android 视图动画。
- 掌握 Android 逐帧动画。
- 掌握 Android 属性动画。

技能目标
- 能创建 Android 的三类动画。

Android 系统提供了丰富的 API，用于实现 UI 的 2D 与 3D 动画，主要可以分为以下三类。

- ◇ 视图动画（View Animation），也叫 Tween（补间）动画，在较老的 Android 版本中就已经提供了，只能用来设置 View 的动画。
- ◇ 逐帧动画（Drawable Animation），其实也可以划分到视图动画一类，专门用来一个一个地显示 Drawable 的图片，就像放幻灯片一样。
- ◇ 属性动画（Property Animation），只对 Android 3.0（API 11）以上的版本有效，这种动画可以设置给任何 Object，包括那些还没有渲染到屏幕上的对象。这种动画是可扩展的，可以让你自定义任何类型和属性的动画。

9.1　创建通风控制系统风扇动画

使用 Android 视图动画来实现通风控制系统的风扇动画。视图动画可以在一个视图容器内执行一系列简单变换，包括淡入淡出（alpha）、位移（translate）、缩放（scale）、旋转（rotate）。例如，如果你有一个 ImageView 对象，可以让其围绕着中心以一定速度进行旋转。视图动画可以通过 XML 或 Android 代码定义，一般建议使用 XML 文件定义，因为它更具可读性和可重用性。下面我们以视图动画的 rotate 为例来让风扇转动起来。

30　创建风扇动画

首先在 activity_main.xml 中为 ImageView 视图定义一个 id。

```
1. <ImageView
2.     android:id="@+id/img_fan"
3.     android:layout_width="0dp"
4.     android:layout_height="wrap_content"
```

```
5. android:layout_weight="1"
6. android:src="@drawable/fan"/>
```

然后在 res/目录下新建 anim 文件夹，在该文件夹下新建 rotate_anim.xml 旋转动画设置文件，内容如下。

```
1. <?xml version="1.0" encoding="utf-8"?>
2. <rotate xmlns:android="http://schemas.android.com/apk/res/android">
3. <rotate
4. android:drawable="@drawable/fan"
5. android:duration="1500"
6. android:fromDegrees="0"
7. android:interpolator="android:anim/linear_interpolator"
8. android:pivotX="50%"
9. android:pivotY="50%"
10. android:repeatCount="-1"
11. android:toDegrees="360"
12. android:visible="true">
13. </rotate>
14. </rotate>
```

第 1 行声明 XML 的版本为 1.0，编码方式为 UTF-8。

第 2 行为 rotate 命名空间定义。

第 3~13 行定义了 rotate 元素。

第 4 行 android:drawable 属性指定了使用 drawable 下的图片 fan。

第 5 行 android:duration 属性指定了动画持续时间，单位为毫秒，这里为 1 500。

第 6 行 android:fromDegrees 属性指定了图片旋转开始角度为 0°，正数代表顺时针度数，负数代表逆时针度数。

第 7 行 android:interpolator 属性指定了使用哪种 interpolator（插值器），主要作用是可以控制动画的变化速率。这里使用线性插值器 linear_interpolator 来实现动画匀速变化。

系统提供了许多已经实现的插值器，如表 9-1 所示。

表 9-1 插值器及功能

XML id 值	功能
accelerate_decelerate_interpolator	动画始末速率较慢，中间加速
accelerate_interpolator	动画开始速率较慢，之后慢慢加速
anticipate_interpolator	开始的时候从后向前甩
anticipate_overshoot_interpolator	类似上面 anticipate_interpolator
bounce_interpolator	动画结束时弹起
cycle_interpolator	循环播放速率改变为正弦曲线
decelerate_interpolator	动画开始快然后慢
linear_interpolator	动画匀速变化
overshoot_interpolator	向前弹出一定值之后回到原来位置

第 8 行 android:pivotX 属性指定了缩放起点 *X* 坐标（数值、百分数、百分数 p，如 50 表示以当前 View 左上角坐标加 50px 为初始点、50%表示以当前 View 的左上角加上当前 View 宽高的 50%作为初始点、50%p 表示以当前 View 的左上角加上父组件宽高的 50%作为初始点）。

第 9 行 android:pivotY 属性指定了缩放起点 *Y* 坐标。

第 10 行 android:repeatCount 属性指定了一个动画的重复次数，int 型。"-1" 表示无限循环；"1" 表示动画在第一次执行完成后重复执行一次，也就是两次；默认为 0，不重复执行。

第 11 行 android:toDegrees 属性指定了旋转结束角度为 360°，正数代表顺时针度数，负数代表逆时针度数。

第 12 行 android:visible 属性指定了图片初始的显示状态为 true（可见），默认为 false（不可见）。

在 MainActivity 活动中使用 AnimationUtils 类加载 rotate_anim.xml 文件。AnimationUtils 类是 Android 系统中的动画工具类，提供了控制 View 对象的一些工具。该类中最常用的方法便是 loadAnimation()方法，该方法用于加载 XML 格式的动画配置文件。在 Android 系统中，除了在代码中设置动画效果外，还可以在 XML 配置文件中设置动画的组合动作，这种方式的适用性更好。

```
private Animation rotate;
rotate=AnimationUtils.loadAnimation(MainActivity.this, R.anim.rotate_anim);
```

在 MainActivity 活动的 onItemSelected()方法中增加打开、关闭通风控制系统代码。

```
1.    @Override
2.    public void onItemSelected(AdapterView<?>parent, View view, int position, long id) {
3.        Context c = getApplicationContext();
4.        String address = smartFactory.getServerAddress();
5.        String projLabel = smartFactory.getProjectLabel();
6.        String controllerId = smartFactory.getVentilationControllerId();
7.        String status = spVentilation.getItemAtPosition(position).toString();
8.        if (cloudHelper.getToken() != "") {
9.            switch (status) {
10.               case "打开":
11.                   cloudHelper.onOff(c, address, projLabel, controllerId, 1);
12.                   ((ImageView)findViewById(R.id.img_fan)).setAnimation(rotate);
13.                   ((ImageView)findViewById(R.id.img_fan)).startAnimation(rotate);
14.                   break;
15.               case "关闭":
16.                   cloudHelper.onOff(c, address, projLabel, controllerId, 0);
17.                   ((ImageView)findViewById(R.id.img_fan)).setAnimation(rotate);
18.                   ((ImageView)findViewById(R.id.img_fan)).clearAnimation();
19.                   break;
20.               case "自动":
21.                   if (Float.parseFloat(tempValue) > smartFactory.getTempThresholdValue()) {
```

```
22.                    cloudHelper.onOff(c, address, projLabel, controllerId,1);
23.                    ((ImageView)findViewById(R.id.img_fan)).setAnimation(rotate);
24.                    ((ImageView)findViewById(R.id.img_fan)).startAnimation(rotate);
25.                } else {
26.                    cloudHelper.onOff(c, address, projLabel,controllerId,0);
27.                    ((ImageView)findViewById(R.id.img_fan)).setAnimation(rotate);
28.                    ((ImageView)findViewById(R.id.img_fan)).clearAnimation();
29.                }
30.                break;
31.            default:
32.                ((ImageView)findViewById(R.id.img_fan)).setAnimation(rotate);
33.                ((ImageView)findViewById(R.id.img_fan)).clearAnimation();
34.                break;
35.        }
36.    }
37. }
```

第 12 行使用 setAnimation()方法指定动画对象。

第 13 行使用 startAnimation()方法启动动画。

第 18 行使用 clearAnimation()方法停止动画。

如果想在代码中设置动画,可以在 MainActivity 活动中增加如下代码。

```
1.  RotateAnimation rotate=new RotateAnimation(
2.      0f,
3.      360f,
4.      Animation.RELATIVE_TO_SELF,
5.      0.5f,
6.      Animation.RELATIVE_TO_SELF,
7.      0.5f);
8.  LinearInterpolator lin=new LinearInterpolator();
9.  rotate.setInterpolator(lin);
10. rotate.setDuration(1500);
11. rotate.setRepeatCount(-1);
12. rotate.setFillAfter(true);
13. rotate.setStartOffset(10);
14. ((ImageView)findViewById(R.id.img_fan)).setAnimation(rotate);
```

第 1~7 行使用 RotateAnimation(float fromDegrees,float toDegrees,int pivotXType,float pivotXValue,int pivotYType,float pivotYValue)方法创建旋转动画对象。第一个参数 fromDegrees 为图片旋转开始角度。第二个参数 toDegrees 为图片旋转结束角度。第三个参数 pivotXType 和第五个参数 pivotYType 为旋转轴点的 X 和 Y 坐标的模式,Animation. RELATIVE_TO_SELF 设置动画组件的左上角为坐标原点(0,0),动画旋转轴的坐标为(view.getWidth*pivotXValue, view.getHeight*pivotYValue),负数向左(X 轴)或上(Y 轴)偏移,正数向右(X 轴)或下

（Y轴）偏移。第四个参数 pivotXValue 和第六个参数 pivotYValue 为旋转轴点 X 和 Y 坐标的相对值。

第 8 行创建一个匀速转动的插值器对象。

第 9 行指定插值器对象。

第 10 行指定动画持续时间为 1 500 毫秒。

第 11 行设置动画重复次数，"-1"为无限循环。

第 12 行指定动画执行完后是否停留在执行完的状态。

第 13 行指定动画执行前的等待时间为 10 毫秒。

第 14 行使用 setAnimation()方法指定动画对象。

31　创建送风动画

9.2　创建空调控制系统送风动画

我们接下来以逐帧动画来模拟空调控制系统打开后空调送风的动作。首先将 ac1.png、ac2.png、ac3.png、ac4.png、ac5.png、ac6.png 这六张图片复制到 res/drawable-xhdpi 目录中，如图 9-1 所示。

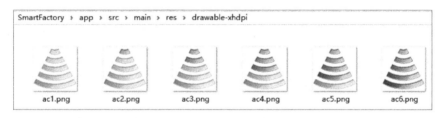

图 9-1　逐帧动画图片

然后在 res/drawable 目录下新建 frame_anim.xml 文件。

```xml
1. <?xml version="1.0" encoding="utf-8"?>
2. <animation-list
3. xmlns:android="http://schemas.android.com/apk/res/android"
4. android:oneshot="false"
5. >
6. <item android:drawable="@drawable/ac1" android:duration="150"></item>
7. <item android:drawable="@drawable/ac2" android:duration="150"></item>
8. <item android:drawable="@drawable/ac3" android:duration="150"></item>
9. <item android:drawable="@drawable/ac4" android:duration="150"></item>
10. <item android:drawable="@drawable/ac5" android:duration="150"></item>
11. <item android:drawable="@drawable/ac6" android:duration="150"></item>
12. </animation-list>
```

第 2 行定义根元素为 animation-list。

第 4 行 android:oneshot 属性设置是否只显示一遍图片，设置为 false 会不停地循环播放动画。

第 6～11 行用 item 元素对动画中的每一个图片进行声明，android:duration 属性指定显示该图片的时间长度为 150 毫秒。

在 MainActivity 中更新 spAc 的 onItemSelected()方法。

```java
1.   @Override
2.   public void onItemSelected(AdapterView<?> parent, View view, int position, long id) {
3.       Context c = getApplicationContext();
4.       String address = smartFactory.getServerAddress();
5.       String projLabel = smartFactory.getProjectLabel();
6.       String controllerId = smartFactory.getAirControllerId();
7.       String status = spAc.getItemAtPosition(position).toString();
8.       ImageView imageView = findViewById(R.id.img_ac);
9.       imageView.setImageResource(R.drawable.frame_anim);
10.      AnimationDrawable ad = (AnimationDrawable)imageView.getDrawable();
11.      if (cloudHelper.getToken() != "") {
12.          switch (status) {
13.              case "打开":
14.                  cloudHelper.onOff(c, address, projLabel, controllerId, 1);
15.                  ad.start();
16.                  break;
17.              case "关闭":
18.                  cloudHelper.onOff(c, address, projLabel, controllerId, 0);
19.                  ad.stop();
20.                  imageView.setImageResource(R.drawable.ac1);
21.                  break;
22.              case "自动":
23.                  if (Float.parseFloat(humValue) > smartFactory.getHumThresholdValue()) {
24.                      cloudHelper.onOff(c, address, projLabel, controllerId,1);
25.                      ad.start();
26.                  } else {
27.                      cloudHelper.onOff(c, address, projLabel, controllerId,0);
28.                      ad.stop();
29.                      imageView.setImageResource(R.drawable.ac1);
30.                  }
31.                  break;
32.              default:
33.                  ad.stop();
34.                  imageView.setImageResource(R.drawable.ac1);
35.                  break;
36.          }
37.      }
38.  }
```

第 9 行设置图片，ImageView 设置图片的方式有很多种，可以在 XML 里面写 android:src="@drawable/xxx"，也可以在 Java 代码里面设置。在 Java 代码里面的设置方式也有多种，方法包括 setImageResource()、setImageDrawable()、setImageBitmap()。在 XML 里面设置实际上和在 Java 代码里面调用 setImageResource()方法是一样的，当然 XML 多了一个解析的过程，放到 Java 代码里调用会稍微好一些（实际上没什么区别）。

setImageResource()方法的参数是 resId，必须是 drawable 目录下的资源。这个方法是在 UI 线程中对图片读取和解析的，所以有可能对一个 Activity 的启动造成延迟。所以如果考虑到这个方面，建议用 setImageDrawable()和 setImageBitmap()方法来代替。

setImageBitmap()方法的参数是 Bitmap，可以解析不同来源的图片再进行设置。实际上 setImageBitmap()方法做的事情就是把 Bitmap 对象封装成 Drawable 对象，然后调用 setImageDrawable()方法来设置图片。因此在编写代码时建议，如果需要频繁调用这个方法的话最好自己封装一个固定的 Drawable 对象，直接调用 setImageDrawable()方法，这样可以减少 Drawable 对象。

setImageDrawable()方法的参数是 Drawable，也是可以接受不同来源的图片，方法中所做的事情就是更新 ImageView 的图片。上面两个方法实际上最后调用的都是 setImageDrawable()方法（setImageResource()方法没有直接调用，不过更新的方法与 setImageDrawable()方法一样）。所以综合来看，setImageDrawable()方法是最省内存、最高效的，如果担心图片过大或图片过多影响内存和加载效率，可以自己解析图片然后通过调用 setImageDrawable()方法进行设置。

第 10 行通过 ImageView 的 getDrawable()方法获得 frame_anim.xml 中的图片生成 AnimationDrawable 对象。

第 15、25 行使用 AnimationDrawable.start()方法打开动画。要注意 AnimationDrawable.start()方法不能直接写在 onClick()、onStart()、onResume()方法里面，这样做是无效的，无法启动动画，只能写在诸如事件监听当中。

第 19、28、33 行使用 AnimationDrawable.stop()方法关闭动画。

从上面的例子可以看出，逐帧动画的动画效果简单、单一。视图动画也存在以下问题。

（1）作用对象局限

视图动画只能够作用在视图上，即只可以对一个 Button、TextView，甚至是 LinearLayout，或者其他继承自 View 的组件进行动画操作，但无法对非视图的对象进行动画操作。

（2）没有改变视图的属性，只是改变视觉效果

视图动画只是改变了视图的视觉效果，而不会真正改变视图的属性。

（3）动画效果单一

视图动画只能实现平移、旋转、缩放、淡入淡出这些简单的动画需求，一旦遇到相对复杂的动画效果，即超出了上述 4 种动画效果，视图动画就无法实现。

9.3 创建照明控制系统灯光动画

下面使用属性动画来实现照明控制系统的打开、关闭。打开时，"关闭图片💡"呈现淡入淡出的效果，时间持续 2s，效果结束后显示"打开图片💡"。

32 创建灯光动画

属性动画会用到两个类，ValueAnimator 类和 ObjectAnimator 类。

ValueAnimator 是整个属性动画机制当中最核心的一个类。属性动画的运行机制是通过不断地对值进行操作来实现的，而初始值和结束值之间的动画过渡就是由 ValueAnimator 这个类来负责计算的。它的内部使用一种时间循环的机制来计算值与值之间的动画过渡，只需

要将初始值和结束值提供给 ValueAnimator，并且告诉它动画所需运行的时长，那么 ValueAnimator 就会自动完成从初始值平滑地过渡到结束值这样的效果。除此之外，ValueAnimator 还负责管理动画的播放次数、播放模式，以及对动画设置监听器等，确实是一个非常重要的类。

下面的例子使用 ValueAnimator 类实现将一个值从 0 平滑过渡到 1，时长 300ms。

```
1. ValueAnimator anim=ValueAnimator.ofFloat(0f, 1f);
2. anim.setDuration(300);
3. anim.start();
```

调用 ValueAnimator 的 ofFloat()方法就可以构建出一个 ValueAnimator 的实例。ofFloat()方法中允许传入多个 float 类型的参数，这里传入 0 和 1 就表示将值从 0 平滑过渡到 1。然后调用 ValueAnimator 的 setDuration()方法来设置动画运行的时长。最后调用 start()方法启动动画。

这只是一个将值从 0 过渡到 1 的动画，还看不到任何界面效果。可以通过 addUpdateListener() 方法来添加一个动画的监听器，在动画执行的过程中会不断地进行回调，只需要在回调方法当中将当前的值取出并打印出来，就可以知道动画有没有真正运行了。

```
1. ValueAnimator anim=ValueAnimator.ofFloat(0f, 1f);
2. anim.setDuration(500);
3. anim.addUpdateListener(new ValueAnimator.AnimatorUpdateListener() {
4.     @Override
5.     public void onAnimationUpdate(ValueAnimator animation) {
6.         float currentValue=(float) animation.getAnimatedValue();
7.         Log.d("TAG", "current value is "+currentValue);
8.     }
9. });
10. anim.start();
```

值从 0 过渡到 1 的动画，执行过程如图 9-2 所示。

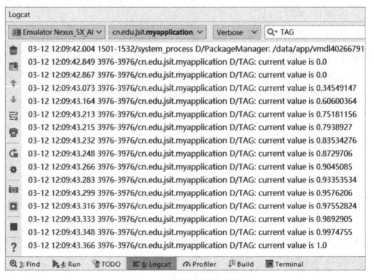

图 9-2 控制台打印输出

另外，ofFloat()方法中是可以传入任意多个参数的，因此还可以构建出更加复杂的动画逻辑。例如，将一个值在 5s 内从 0 过渡到 5，再过渡到 3，再过渡到 10，就可以这样写：

```
1. ValueAnimator anim=ValueAnimator.ofFloat(0f, 5f, 3f, 10f);
2. anim.setDuration(5000);
3. anim.start();
```

如果不需要小数位数的动画过渡，只需要调用 ValueAnimator 的 ofInt()方法就可以了，如下所示。

```
ValueAnimator anim=ValueAnimator.ofInt(0, 100);
```

ValueAnimator 有个缺点，就是只能用数值对动画计算。为了能让动画直接与对应组件相关联，以使我们从监听动画过程中解放出来，在 ValueAnimator 的基础上，又派生了一个类 ObjectAnimator。由于 ObjectAnimator 是派生自 ValueAnimator 的，所以 ValueAnimator 中能使用的方法，在 ObjectAnimator 中都可以正常使用。

下面利用 ObjectAnimator 重写的 ofFloat()方法实现一个改变图片视图透明度的动画。

```
1. ObjectAnimator objectAnimator=ObjectAnimator.ofFloat(imgLight,"alpha", 0.5f);
2. ObjectAnimator.setDuration(1000);
3. objectAnimator.start();
```

offFloat()方法中第一个参数用于指定这个动画要操作的是哪个组件；第二个参数用于指定这个动画要操作这个组件的哪个属性，只需要改变 ofFloat()的第二个参数值就可以实现对应的动画；第三个参数是可变长参数，这就跟 ValueAnimator 中的可变长参数的意义一样了，就是指这个属性值是从哪儿变到哪儿。

在 MainActivity 活动中通过代码实现动画。

```
1.  final ImageView imgLight = findViewById(R.id.img_light);
2.  final ObjectAnimator oa = ObjectAnimator.ofFloat(imgLight,"alpha",1f,0f,1f).setDuration(2000);
3.  oa.addListener(new AnimatorListenerAdapter() {
4.      @Override
5.      public void onAnimationEnd(Animator animation) {
6.          super.onAnimationEnd(animation);
7.          imgLight.setImageResource(R.drawable.light_on);
8.      }
9.  });
10. spLight = findViewById(R.id.sp_light_control);
11. spLight.setAdapter(adapter);
12. spLight.setOnItemSelectedListener(new AdapterView.OnItemSelectedListener() {
13.     @Override
14.     public void onItemSelected(AdapterView<?> parent, View view, int position, long id) {
15.         Context c = getApplicationContext();
16.         String address = smartFactory.getServerAddress();
```

```
17.         String projLabel = smartFactory.getProjectLabel();
18.         String controllerId = smartFactory.getLightControllerId();
19.         String status = spLight.getItemAtPosition(position).toString();
20.         if (cloudHelper.getToken() != "") {
21.             switch (status) {
22.                 case "打开":
23.                     cloudHelper.onOff(c, address, projLabel, controllerId, 1);
24.                     oa.start();
25.                     break;
26.                 case "关闭":
27.                     cloudHelper.onOff(c, address, projLabel, controllerId, 0);
28.                     imgLight.setImageResource(R.drawable.light_off);
29.                     break;
30.                 case "自动":
31.                     if (Float.parseFloat(lightValue) > smartFactory.getLightThresholdValue()) {
32.                         cloudHelper.onOff(c, address, projLabel, controllerId, 1);
33.                         oa.start();
34.                     } else {
35.                         cloudHelper.onOff(c, address, projLabel, controllerId, 0);
36.                         imgLight.setImageResource(R.drawable.light_off);
37.                     }
38.                     break;
39.                 default:
40.                     imgLight.setImageResource(R.drawable.light_off);
41.                     break;
42.             }
43.         }
44.     }
45.
46.     @Override
47.     public void onNothingSelected(AdapterView<?> parent) {
48.
49.     }
50. });
51. spLight.setSelection(1,true);
```

第 2 行创建了一个透明度从 1 到 0，再从 0 到 1 的渐变动画。时间长度为 2 000ms。

第 3~9 行实现了添加一个监听器。在很多时候，我们希望可以监听到动画的各种事件，如动画何时开始，何时结束，然后在开始或结束时执行一些逻辑处理。这个功能是完全可以实现的，Animator 类中提供了一个 addListener()方法，这个方法接收一个 AnimatorListener，我们只需要实现 AnimatorListener 就可以监听动画的各种事件了。前面已经讲过，ObjectAnimator 是继承自 ValueAnimator 的，而 ValueAnimator 又是继承自 Animator 的，因此不管是 ValueAnimator 还是 ObjectAnimator 都可以使用 addListener()这个方法。另外
AnimatorSet 也是继承自 Animator 的，因此 addListener()这个方法算是个通用的方法。

```
1.  objectAnimatior.addListener(new AnimatorListener() {
2.      @Override
3.          public void onAnimationStart(Animator animation) {
4.      }
5.
6.      @Override
7.          public void onAnimationRepeat(Animator animation) {
8.      }
9.
10.     @Override
11.         public void onAnimationEnd(Animator animation) {
12.     }
13.
14.     @Override
15.         public void onAnimationCancel(Animator animation) {
16.     }
17. });
```

可以看到，需要实现接口中的四个方法，onAnimationStart()方法会在动画开始的时候调用，onAnimationRepeat()方法会在动画重复执行的时候调用，onAnimationEnd()方法会在动画结束的时候调用，onAnimationCancel()方法会在动画被取消的时候调用。

但是很多时候并不需要监听那么多个事件，可能只要监听动画结束这一个事件，那么每次都要将四个接口全部实现一遍就显得非常烦琐。为此 Android 提供了一个适配器类，叫作 AnimatorListenerAdapter，使用这个类就可以解决实现接口烦琐的问题了。

第 7 行中当监听动画结束时，imgLight 视图显示图片 light_on。

下面使用代码来编写所有的属性动画功能，这也是最常用的一种做法。当然也可以通过 XML 来编写动画。首先在 res/目录下创建一个 animator 文件夹，然后新建一个 anim.xml 文件，如下所示。

```
1.  <set xmlns:android="http://schemas.android.com/apk/res/android"
2.      android:ordering="sequentially">
3.      <set android:ordering="sequentially">
4.      <objectAnimator
5.          android:duration="1000"
6.          android:propertyName="alpha"
7.          android:valueFrom="1"
8.          android:valueTo="0"
9.          android:valueType="floatType">
10.     </objectAnimator>
11.     <objectAnimator
12.         android:duration="1000"
13.         android:propertyName="alpha"
```

```
14.            android:valueFrom="0"
15.            android:valueTo="1"
16.            android:valueType="floatType">
17.        </objectAnimator>
18.    </set>
19. </set>
```

第 4~10 行将一个视图的 alpha 属性从 1 变成 0，时间为 1 000ms。
第 11~17 行将一个视图的 alpha 属性从 0 变成 1，时间为 1 000ms。
修改 MainActivity 活动如下。

```
1. Animator animator=AnimatorInflater.loadAnimator(this, R.animator.anim);
2. animator.setTarget(imageView);
3. animator.start();
4. animator.addListener(new AnimatorListenerAdapter() {
5.     @Override
6.     public void onAnimationEnd(Animator animation) {
7.         imageView.setImageResource(R.drawable.light_on);
8.     }
9. });
```

第 1 行调用 AnimatorInflater 的 loadAnimator()方法将 XML 动画文件加载进来。
第 2 行调用 setTarget()方法将这个动画设置到 imageView 对象上面。
第 3 行调用 start()方法启动动画。

任务 10　绘制传感器数据折线图

任务概述

所有的应用都需要存储数据。在 SmartFactory 应用中，需要保存温度、湿度、光照度数据，用户可以通过折线图查看这些历史数据，如图 10-1 所示。

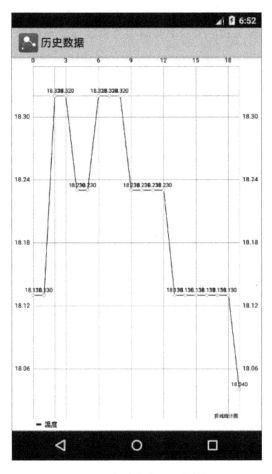

图 10-1　查看传感器历史数据

知识目标
- 掌握 SQLite 数据库。
- 掌握 MPAndroidChart 图标库。

技能目标
- 能通过 SQLite 数据库来保存、查询数据。
- 能使用 MPAndroidChart 图表库来绘制数据图。

10.1 使用 SQLite 数据库保存数据

33 使用 SQLite 数据库和创建帮助器

Android 存储数据的主要方法就是使用 SQLite 数据库。SQLite 数据库具有以下特点。

（1）轻量级的数据库管理系统

大多数数据库管理系统都需要一个特殊的数据库服务器进程才能工作。SQLite 并不需要这样的一个服务器进程，它使用单个文件存储数据，因此不会占用任何处理器时间。这对于移动设备非常重要，可以减少电池电量的消耗。

（2）针对单个用户优化

只是当前应用与数据库交互，所以不必用用户名和口令来验证。

（3）稳定且速度快

SQLite 数据库极其稳定，可以处理数据库事务，当出现数据库故障时，SQLite 可以回滚数据。SQLite 读写数据的代码是用优化的 C 语言编写的，所以执行速度快，可以减少处理器功率消耗。

Android 会自动为各个应用创建一个文件夹来存储应用的数据库。为 SmartFactory 应用创建数据库时，它会存储在/data/data/cn.edu.jsit.smartfactory/databases 文件夹中。例如，我们会在本任务中创建数据库“save_data”，如图 10-2 所示。

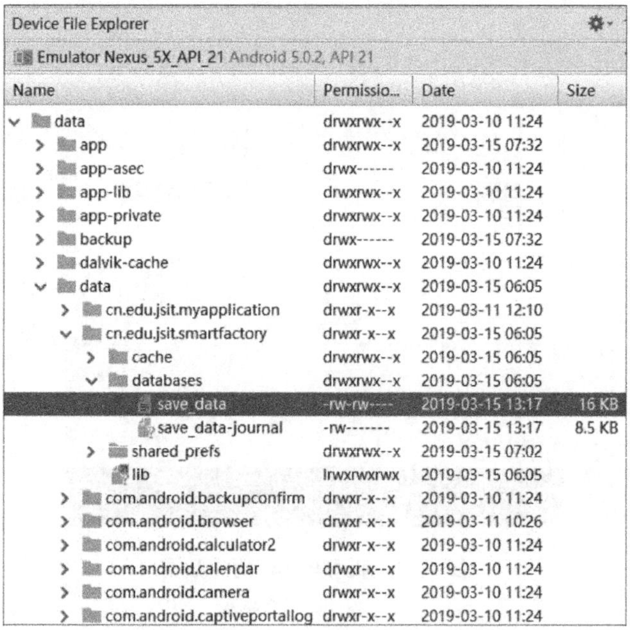

图 10-2 数据库目录

在这个文件夹中可以存储多个数据库。每个数据库由两个文件组成。第一个文件是数据库文件，与数据库同名，如“save_data”。这是主 SQLite 数据库文件，所有数据都存放在这个文件中。第二个文件是 journal 文件。它也与数据库同名，另外有一个“-journal”后缀，如"save_data-journal"。这个 journal 文件包含对数据库的所有修改，如果出现问题，Android 会使用这个 journal 文件撤销（或回滚）最近的修改。

Android 使用了一组类，可以利用这些类来管理 SQLite 数据库。

（1）SQLite 帮助器

可以继承 SQLiteOpenHelper 类创建一个 SQLiteOpenHelper 子类（SQLite 帮助器）来创建和管理数据库。

（2）SQLiteDatabase

可以利用 SQLiteDatabase 类访问数据库。在 Android 中使用 SQLiteDatabase 的静态方法 openOrCreateDatabase()打开或创建一个数据库。就像 JDBC 中的 SQLConnection。

（3）Cursor（游标）

可以利用 Cursor 读写数据库。就像 JDBC 中的 ResultSet。

10.2 创建 SQLite 帮助器

可以写一个类来继承 SQLiteOpenHelper 类来创建 SQLite 帮助器。在 SQLite 帮助器子类中必须重写 onCreate()和 onUpgrade()两个方法。因为第一次在设备上创建数据库时会调用 onCreate()方法。在这个方法中实现数据库的创建和数据库表的创建。当数据库要升级时会调用 onUpgrade()方法。例如，当需要在发布数据库之后对数据库中的表进行修改，就要在这个方法中完成。

```
1.  package cn.edu.jsit.smartfactory.tools;
2.
3.  import android.content.ContentValues;
4.  import android.content.Context;
5.  import android.database.Cursor;
6.  import android.database.sqlite.SQLiteDatabase;
7.  import android.database.sqlite.SQLiteOpenHelper;
8.  import java.util.ArrayList;
9.  import java.util.List;
10.
11. public class DataBaseHelper extends SQLiteOpenHelper {
12.     private SQLiteDatabase db;
13.     private static final String DB_NAME="smartfactory";
14.     private static final int DB_VERSION=1;
15.
16.     public DataBaseHelper(Context context) {
17.         super(context, DB_NAME, null, DB_VERSION);
18.     }
19.
20.     @Override
21.     public void onCreate(SQLiteDatabase db) {
22.         db.execSQL("create table data(id integer primary key,"
23.             +"temperature text,"
```

```java
24.            +"humidity text,"
25.            +"light text)");
26.         db.beginTransaction();
27.         try {
28.            db.execSQL(" create trigger trigger_delete_top "
29.               +" AFTER insert on data"
30.               +" BEGIN delete from data "
31.               +" where (select count(id) from data) > 20 "
32.               +" and id in (select id from data  order by id asc "
33.               +" limit  (select count(id)-20 from data));"
34.               +" END;");
35.            db.setTransactionSuccessful();
36.         } catch(Exception e) {
37.            e.printStackTrace();
38.         } finally {
39.            db.endTransaction();
40.         }
41.      }
42.
43.      private void read(Context context) {
44.         DataBaseHelper dataBaseHelper=new DataBaseHelper(context);
45.         db=dataBaseHelper.getReadableDatabase();
46.      }
47.
48.      private void write(Context context) {
49.         DataBaseHelper dataBaseHelper=new DataBaseHelper(context);
50.         db=dataBaseHelper.getWritableDatabase();
51.      }
52.
53.      public void insert(Context context, String temp, String hum, String light) {
54.         write(context);
55.         ContentValues cv=new ContentValues();
56.         cv.put("temperature", temp);
57.         cv.put("humidity", hum);
58.         cv.put("light", light);
59.         db.insert("data", null, cv);
60.         close();
61.      }
62.      public List<Float> search(Context context,String type) {
63.         read(context);
64.         List<Float> data=new   ArrayList<Float>();
65.         Cursor c=db.rawQuery("select * from data order by id desc", null);
66.         while (c.moveToNext()) {
67.            float s=0;
68.            switch (type) {
```

```
69.                 case "温度":
70.                     s=Float.parseFloat(c.getString(1));
71.                     break;
72.                 case "湿度":
73.                     s=Float.parseFloat(c.getString(2));
74.                     break;
75.                 case "光照":
76.                     s=Float.parseFloat(c.getString(3));
77.                     break;
78.             }
79.             data.add(s);
80.         }
81.         close();
82.         return data;
83.     }
84.
85.     public void close() {
86.         if (db != null)
87.             db.close();
88.     }
89.
90.     @Override
91.     public void onUpgrade(SQLiteDatabase db,int oldVersion, int newVersion) {
92.
93.     }
94. }
```

第 11 行创建 SQLite 帮助器 DataBaseHelper（继承 SQLiteOpenHelper）。

第 12 行定义 SQLiteDatabase 引用 SQL。

第 13 行定义数据库名称。

第 14 行定义数据库版本。

第 16~18 行定义了构造函数。第 17 行调用 SQLiteOpenHelper 父类的构造函数，并传入数据名称和版本。DataBaseHelper 负责在第一次需要使用数据库时创建 SQLite 数据。首先在设备上创建一个空的数据库，然后调用 DataBaseHelper 的 onCreate()方法，向 onCreate()方法传入一个 SQLiteDatabase 对象作为参数。可以使用这个参数用下面的方法运行 SQL 命令。

```
SQLiteDatabase.execSQL(String sql);
```

第 20~41 行重写了 onCreate()方法。第 22~25 行创建了一个空的 data 表，如表 10-1 所示。

表 10-1 data 表

id	temperture	humility	light

第 26~40 行创建了一个触发器 trigger_delete_top，用于限制数据库仅保存最新的 20 行数据。

第 43~46 行定义了 read()方法。

第 48~51 行定义了 write()方法。

第 53~60 行定义了 insert()方法。如果需要在一个 SQLite 表中插入数据，可以使用 SQLiteDatabase 的 insert()方法。这个方法允许在数据库中插入数据，一旦插入完成，会返回所插入记录的 ID。如果用这个方法不能插入这条记录，则返回值-1。

第 55 行创建了一个 ContentValues 对象。要使用 insert()方法，需要指定想要将数据插入哪个表，以及所插入的值。指定想要插入的值时，需要创建一个 ContentValues 对象。

第 56~58 行用 ContentValues 对象的 put()方法为这个对象增加名/值数据对。

第 59 行用 SQLiteDatabase 的 insert()方法在 data 表中插入值。insert()方法的第一个参数为表名；第二个参数设置为 null，之所以为 null，是因为 ContentValues 对象可以为空，你可能希望在表中插入一个空行，而 SQLite 不允许插入空行，除非至少指定了一个列的列名，利用这个参数就可以指定其中的一个列；第三个参数为 ContentValues 对象。

第 62~83 行定义了 search()方法，用于查询数据。

第 63 行调用 read()方法，在 read()方法中通过 getWritableDatabase()获取一个用于操作数据库的 SQLiteDatabase 实例。

第 64 行创建一个 float 类型的 ArrayList 对象。

第 65 行创建了一个 Cursor（游标），可以利用游标访问数据库信息，在数据库记录之间导航。可以使用 SQLiteDatabase 的 rawQuery()方法建立一个查询，rawQuery()方法返回一个类型为 Cursor 的对象，活动可以使用这个对象来访问数据库。rawQuery()方法几乎支持所有的 SQL 查询语句，第一个参数为 SQL 查询字符串，第二个参数为字符串数组（由查询的数据转换得到）。例如：

```
db.rawQuery("select name from data where id=?", new String[]{"1"});
```

也可以使用 SQLiteDatabase 的 query()方法建立一个查询，query()方法是 Android 自己封装的查询 API。query()方法和 rawQuery()一样都是返回一个对象。query()方法有 7 个参数，如下所示。

```
public Cursor query(String table,
        String[] columns,
        String selection,
        String[] selectionArgs,
        String groupBy,
        String having,
        String orderBy)
```

第一个参数是访问的表名，第二个参数是列名，第三、四个是查询条件参数，第五、六个是使用聚合函数时需要的参数，第七个是排序参数。

两种方法没有本质区别，最后调用的都是同一个方法 rawQueryWithFactory()。

第 66~79 行使用 Cursor 的 moveToNext()方法移动到下一条记录，如果成功会返回 true，如果失败则返回 false。根据数据类型（温度、湿度、光照度）生成 data。

第 81 行使用 SQLiteDatabase 的 close()方法关闭数据库。

第 90～93 行重写了 SQLiteDatabase 的 onUpgrade()方法，这里为空。这个方法用于升级数据库，第一个参数是 SQLiteDatabase 对象，第二个参数是当前数据库版本，第三个参数是新版本。要升级数据库，新版本必须高于老版本，如可以将第 14 行中的 1 改成 2。相应地也可以使用 onDowngrade()方法降级数据库。

在 MainActivity 中更新 loadCloudData()方法，在从云平台获取数据后直接保存到 SQLite 数据库。如下面代码的第 6、7 行所示，使用了 insert()方法，插入数据。

```
1.  new Timer().schedule(new TimerTask() {
2.      @Override
3.      public void run() {
4.          if (cloudHelper.getToken() != "") {
5.              ...
6.              if (!((tempValue == null) && (humValue == null) && (lightValue==null)))
7.                  databaseHelper.insert(MainActivity.this, tempValue, humValue, lightValue);
8.                  handler.sendEmptyMessage(1);
9.          }
10.     }
11. }, 0, 5000);
```

10.3 使用 MPAndroidChart 来绘制传感器数据折线图

34 绘制传感器数据折线图

MPAndroidChart 是一款基于 Android 的开源图表库，MPAndroidChart 不仅可以在 Android 设备上绘制各种统计图表，而且可以对图表进行拖动和缩放操作，应用起来非常灵活。MPAndroidChart 显得更为轻巧和简单，拥有常用的图表类型，包括线型图、饼图、柱状图和散点图。

10.3.1 导入 MPAndroidChart 图表库

要使用 MPAndroidChart 需要在 app 目录下的 build.gradle 中添加 MPandroidChart 依赖。在 Project 目录下的 build.gradle 文件中添加如下代码。

```
1.  allprojects {
2.      repositories {
3.          jcenter()
4.          maven { url "https://jitpack.io" }
5.      }
6.  }
```

在 app 目录下的 build.gradle 文件中添加依赖。

```
implementation 'com.github.PhilJay:MPAndroidChart:v3.1.0-alpha'
```

使用 MPAndroidChart 的步骤如下。
1）定义布局指定使用哪种图表。

2）绑定 View 设置 LineChart 显示属性。
3）绑定图表数据。
4）设置 X、Y 轴的显示效果。
5）设置图表交互。
6）设置图例。
7）设置 MarkView 提示，单击交点的小提示窗。
8）设置动画。

下面新建 activity_chart.xml 布局文件。该文件使用线性布局，其中定义了一个折线图表。另外还有 BarChar、HorizontalBarChart、RadarChart、PieChart 等常用的图表。

```xml
1. <?xml version="1.0" encoding="utf-8"?>
2. <LinearLayout xmlns:android="http://schemas.android.com/apk/res/android"
3.     android:layout_width="match_parent"
4.     android:layout_height="match_parent">
5.     <com.github.mikephil.charting.charts.LineChart
6.         android:id="@+id/lineChart"
7.         android:layout_width="match_parent"
8.         android:layout_height="match_parent"
9.         />
10. </LinearLayout>
```

10.3.2 创建活动 DataChartActivity

下面创建 DataChartActivity 活动来显示各类传感器数据的折线图。

```java
1.  package cn.edu.jsit.smartfactory;
2.
3.  import android.app.Activity;
4.  import android.content.Intent;
5.  import android.graphics.Color;
6.  import android.graphics.Typeface;
7.  import android.os.Bundle;
8.  import android.support.annotation.Nullable;
9.  import android.util.Log;
10. import com.github.mikephil.charting.charts.LineChart;
11. import com.github.mikephil.charting.components.Description;
12. import com.github.mikephil.charting.components.Legend;
13. import com.github.mikephil.charting.data.Entry;
14. import com.github.mikephil.charting.data.LineData;
15. import com.github.mikephil.charting.data.LineDataSet;
16. import com.github.mikephil.charting.utils.ColorTemplate;
17. import java.util.ArrayList;
18. import java.util.List;
19. import cn.edu.jsit.smartfactory.tools.DataBaseHelper;
20.
21. public class DataChartActivity extends Activity {
```

```java
22.     DataBaseHelper dataBaseHelper;
23.     private LineChart lineChart;
24.     protected Typeface typeface;
25.     private ArrayList<Entry> sensorData=new ArrayList<>();
26.     private String dataType;
27.
28.     @Override
29.     protected void onCreate(@Nullable Bundle savedInstanceState) {
30.         super.onCreate(savedInstanceState);
31.         setContentView(R.layout.activity_chart);
32.         Intent intent=getIntent();
33.         dataType=intent.getStringExtra("type");
34.         dataBaseHelper=new DataBaseHelper(this);
35.         lineChart=findViewById(R.id.lineChart);
36.         Description description=new Description();
37.         description.setText("折线统计图");
38.         lineChart.setDescription(description);
39.         lineChart.setDrawGridBackground(false);
40.         lineChart.setBackgroundColor(Color.WHITE);
41.         lineChart.setData(getLineData());
42.         lineChart.setScaleEnabled(false);        //XY轴禁止缩放
43.         Legend legend=lineChart.getLegend();     //设置图例样式
44.         legend.setForm(Legend.LegendForm.LINE);
45.         legend.setTypeface(typeface);
46.         legend.setTextSize(11f);
47.         legend.setTextColor(Color.BLACK);
48.         legend.setVerticalAlignment(Legend.LegendVerticalAlignment.BOTTOM);
49.         legend.setHorizontalAlignment(Legend.LegendHorizontalAlignment.LEFT);
50.         legend.setOrientation(Legend.LegendOrientation.HORIZONTAL);
51.         legend.setDrawInside(false);
52.         lineChart.animateX(3000);//数据显示动画
53.     }
54.
55.     private LineData getLineData() {
56.         getChartData(20);
57.         LineDataSet dataSet=new LineDataSet(sensorData,dataType);
58.         dataSet.setColor(Color.BLUE);
59.         dataSet.setFillColor(ColorTemplate.getHoloBlue());
60.         dataSet.setHighLightColor(Color.rgb(244, 117, 117));
61.         dataSet.setDrawCircleHole(true);
62.         LineData data=new LineData(dataSet);
63.         data.setValueTextColor(Color.BLACK);
64.         data.setValueTextSize(9f);
65.         return data;
66.     }
67.
```

```
68.     private void getChartData(int count) {
69.         List<Float> lists;
70.         lists=dataBaseHelper.search(DataChartActivity.this,dataType);
71.         if(count>lists.size()) {
72.             for(int i=0; i<lists.size(); i++) {
73.                 Entry tempEntry=new Entry(i, lists.get(i));
74.                 sensorData.add(tempEntry);
75.             }
76.         } else {
77.             for(int i=0; i<count; i++) {
78.                 Entry tempEntry=new Entry(i, lists.get(i));
79.                 sensorData.add(tempEntry);
80.             }
81.         }
82.     }
83. }
84. }
```

第 36 行创建描述信息对象 description。

第 37 行设置图表显示名称。

第 38 行设置图表描述信息。

第 39 行取消图表网格背景。

第 40 行设置图表背景颜色。

第 41 行调用 getLineData()方法绑定图表数据。

第 56 行通过 getChartData(20)方法获取最近的 20 个数据。

第 57~61 行设置数据显示格式。

第 62~64 行设置显示值的颜色和文字大小。

第 70~82 行通过调用 dataBaseHelper 的 search()方法从 SQLite 表 data 中获取当前类型的数据，如果少于 20 个，则根据实际数量将数据添加到 sensorData 中，如果大于或等于 20 个则取添加最新的 20 个。

在 MainActivity 中为 tvTemp、tvHum、tvLight 添加单击事件。通过 view.getId()方法获取视图的 id，根据 id 的不同，为意图增加数据类型标识"温度""湿度""光照"。

更新 MainActivity，为三个 TextView 添加监听器，并重写 OnClickListener 接口中的 onClick()方法。

```
1. ...
2. tvTemp.setOnClickListener(this);
3. tvHum.setOnClickListener(this);
4. tvLight.setOnClickListener(this);
5. ...
6. @Override
7. public void onClick(View view) {
8.     Intent intent=new Intent(MainActivity.this,DataChartActivity.class);
9.     switch(view.getId()) {
```

```
10.         case R.id.tv_temp_value:
11.             intent.putExtra("type","温度");
12.             startActivity(intent);
13.             break;
14.         case R.id.tv_hum_value:
15.             intent.putExtra("type","湿度");
16.             startActivity(intent);
17.             break;
18.         case R.id.tv_light_value:
19.             intent.putExtra("type","光照");
20.             startActivity(intent);
21.             break;
22.     }
23. }
```

第 8 行创建意图实现 MainActivity 到 DataChartActivity 的跳转。

第 9 行通过视图的 getId()方法获取 id 值。

第 10～21 行根据 id 值来为意图添加数据类型信息（在 DataChartActivity 中会根据该数据类型信息从 SQLite 数据库表中获取相应的数据），并启动意图。

三种类型传感器历史数据折线图中的温度折线图如图 10-3 所示，湿度折线图如图 10-4 所示，光照度折线图如图 10-5 所示。

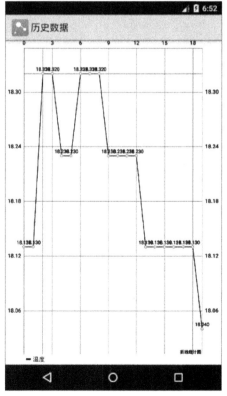
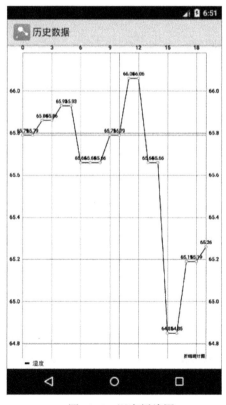

图 10-3　温度折线图　　　　　　　　　　图 10-4　湿度折线图

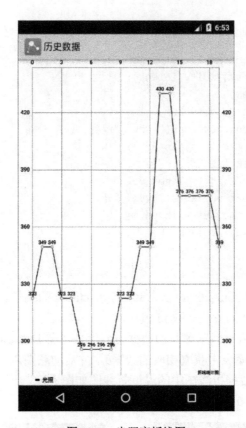

图 10-5 光照度折线图

任务 11　存储报警信息至服务器并创建警报数据界面

任务概述

当禁入区检测到有人时，将有人闯入的报警信息存储至服务器。主界面的禁入区域监控区域出现"有人闯入"的提示信息如图 11-1 所示。单击"有人闯入"的提示信息组件，跳转到警报数据界面，警报数据界面以列表框的形式显示所有报警信息，如图 11-2 所示。

图 11-1　禁入区域监控信息　　　　图 11-2　警报数据界面

知识目标

- 掌握 WebService 的访问方法。
- 掌握 ListView 组件。
- 掌握自定义 Adapter（适配器）。

技能目标

- 能通过访问 WebService 将信息保存至服务器。
- 能通过访问 WebService 从服务器读取信息。
- 能创建自定义适配器为 ListView 组件绑定数据并显示信息。

35　创建和
部署用户管理
WebService

11.1　创建和部署 WebService

这里的数据存储和访问通过调用服务器端 WebService 来实现，采用客户端—服务器端方式。这种方式的数据交互如图 11-3 所示。

图 11-3　数据交互模式

客户端（Client）定时将人体红外传感器检测到的信息发送至服务器端（Server）进行保存，当客户端需要查看警报信息时，先发送请求给服务器端，服务器端提供警报信息给客户端。要与服务器端进行数据交互，则首先需要创建 WebService 服务。WebService 服务是一种跨编程语言和跨操作系统平台的远程调用技术，即跨平台远程调用技术。XML、SOAP 和 WSDL 就是构成 WebService 平台的三大技术。

1．XML

WebService 采用 HTTP 在客户端和服务器端之间传输数据。WebService 使用 XML 来封装数据，XML 的主要优点在于它是跨平台的。

2．SOAP

WebService 通过 HTTP 发送请求和接收结果时，发送的请求内容和结果内容都采用 XML 格式封装，并增加了一些特定的 HTTP 消息头，以说明 HTTP 消息的内容格式，这些特定的 HTTP 消息头和 XML 内容格式就是 SOAP 规定的。SOAP 即简单对象访问协议（Simple Object Access Protocal），是一种简单的基于 XML 的协议，它使应用程序通过 HTTP 来交换信息，简单理解为 SOAP=HTTP+XML。SOAP 不是 WebService 的专有协议，其他应用协议也可使用 SOAP 传输数据。

3．WSDL

WebService 服务器端首先要通过一个 WSDL 文件来说明自己有什么服务可以对外调用。简单地说，WSDL 就像是一个说明书，用于描述 WebService 及其方法、参数和返回值。WSDL 文件保存在 Web 服务器上，通过一个 URL 地址就可以访问。客户端调用一个 WebService 服务之前，要知道该服务的 WSDL 文件的地址。WebService 服务提供商可以通过两种方式来暴露它的 WSDL 文件地址。

（1）注册到 UDDI 服务器，以便被人查找。

（2）直接告诉客户端调用者。

WebService 交互的过程就是，服务器端放在 Web 服务器（如 IIS）中，客户端生成的

SOAP 请求会被嵌入在一个 HTTP POST 请求中，发送到 Web 服务器。Web 服务器再把这些请求转发给服务器端的请求处理器。请求处理器的作用在于，解析收到的 SOAP 请求，调用服务器端，然后再生成相应的 SOAP 应答。Web 服务器得到 SOAP 应答后，会再通过 HTTP 应答的方式把它送回到客户端。

11.1.1 创建 WebService

本任务中的 WebService 服务使用 C#语言编写，开发环境为 Visual Studio 2017（如果对 C#不了解，可以忽略该内容，不会影响本书的阅读）。具体创建代码如下所示。

```csharp
1.  using System;
2.  using System.Collections.Generic;
3.  using System.Linq;
4.  using System.Web;
5.  using System.Web.Services;
6.  using System.Data;
7.  using System.Data.Sql;
8.  using System.Data.SqlClient;
9.  using Newtonsoft.Json;
10.
11. namespace Test {
12.     [WebService(Namespace="http://tempuri.org/")]
13.     [WebServiceBinding(ConformsTo=WsiProfiles.BasicProfile1_1)]
14.     [System.ComponentModel.ToolboxItem(false)]
15.     public class Test:System.Web.Services.WebService {
16.         public static string strcon="data source=.;
17.          initial catalog=Test;user id=sa;password=123456";
18.
19.         [WebMethod(Description="存储报警信息 参数(info:报警信息)")]
20.         public int SaveInfo(string info) {
21.             int number=0;
22.             int id=0;
23.             using(SqlConnection con=new SqlConnection(strcon)) {
24.                 con.Open();
25.                 SqlCommand cmd=new SqlCommand(
26. "insert into info values (@info)", con);
27.                 cmd.Parameters.Add("info", SqlDbType.NVarChar).Value=info;
28.                 number=cmd.ExecuteNonQuery();
29.
30.                 SqlCommand cmd0=new SqlCommand(
31. "select COUNT(*) from info", con);
32.                 if(Convert.ToInt32(cmd0.ExecuteScalar())>50) {
33.                     SqlCommand cmd1=new SqlCommand(
34. "select top 1 * from info order by id asc", con);
35.                     SqlDataReader re=cmd1.ExecuteReader();
```

```
36.                    if(re.Read()) {
37.                        id=Convert.ToInt32(re[0]);
38.                    }
39.                    re.Close();
40.                    SqlCommand cmd2=new SqlCommand(
41. "delete from info where id='"+id+"'", con);
42.                    number=cmd2.ExecuteNonQuery();
43.                }
44.            }
45.            return number;
46.        }
47.
48.        [WebMethod(Description="获取数据库所有信息")]
49.        public string GetInfo() {
50.            string result="";
51.            List<Info> results=new List<Info>();
52.            using(SqlConnection con=new SqlConnection(strcon)) {
53.                con.Open();
54.                SqlCommand cmd=new SqlCommand("Select * from info", con);
55.                SqlDataReader re=cmd.ExecuteReader();
56.                while(re.Read()) {
57.                    Info info=new Info();
58.                    info.ID=re[0].ToString();
59.                    info.Message=re[1].ToString();
60.                    results.Add(info);
61.                }
62.                con.Close();
63.            }
64.            result=JsonConvert.SerializeObject(results);
65.            return result;
66.        }
67.    }
68.
69.    public class Info {
70.        public string ID {
71.            set;
72.            get;
73.        }
74.        public string Message {
75.            get;
76.            set;
77.        }
78.    }
79. }
```

第 19~46 行创建了 SaveInfo()方法，用于保存报警信息。

第 48~66 行创建了 GetInfo()方法，用于获取报警信息。

11.1.2 部署 WebService

创建好服务器端的 WebService 后(也可从本任务资源包中获取),需要在 IIS 中部署 WebService,如图 11-4 所示。

图 11-4 部署 WebService

从任务资源包中找到数据库文件"Test",将其附加到 SQL Server 数据库中,如图 11-5 所示。

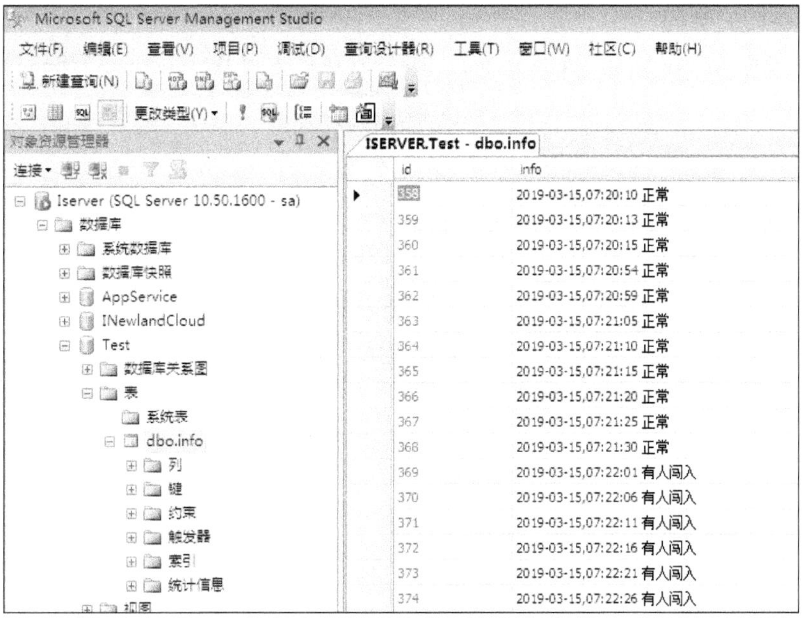

图 11-5 附加到数据库

11.2 创建 WebServiceHelper 类

37 创建 WebServiceHelper 类

在 Android SDK 中并没有提供调用 WebService 的库，因此需要使用第三方类库 Ksoap2 来实现 WebService 的编程，其主要作用是进行数据的封装请求和解析。本任务中需要用到 Ksoap2.jar 库，在工程的 libs 目录下导入 Ksoap2.jar（从本任务资源包中获取），具体导入过程可以参照 7.1 节，如图 11-6 所示。

图 11-6 导入 Ksoap2

使用 Ksoap2 创建 WebServiceHelper 类封装对 WebService 的访问。

WebServiceHelper 类中主要包含两个方法，分别是 SaveInfo()和 GetInfo()。SaveInfo()方法的作用是保存数据至服务器，GetInfo()方法的作用是从服务器获取数据。无论是从 WebService 保存数据还是获取数据都需要访问 WebService。另外，定义了一个回调接口 Callback 返回获取的数据。

此处我们使用 SOAP 访问 WebService。基于 SOAP 访问 WebService 方法的代码如下：

```
1. package cn.edu.jsit.smartfactory.tools;
2.
3. import android.util.Log;
4. import org.ksoap2.SoapEnvelope;
5. import org.ksoap2.serialization.SoapObject;
6. import org.ksoap2.serialization.SoapSerializationEnvelope;
7. import org.ksoap2.transport.HttpTransportSE;
8. import org.xmlpull.v1.XmlPullParserException;
9. import java.io.IOException;
10.
11. public class WebServiceHelper {
12.
13.     public interface Callback {
14.         void call(String s);
15.     }
16.
```

```
17.      public static void SaveInfo(final String msg) {
18.          new Thread() {
19.              public void run() {
20.                  SoapSerializationEnvelope envelope=new SoapSerializationEnvelope(SoapEnvelope.VER10);
21.                  SoapObject rpc=new SoapObject("http://tempuri.org/", "SaveInfo");
22.                  rpc.addProperty("info", msg);
23.                  envelope.bodyOut=rpc;
24.                  envelope.dotNet=true;
25.                  envelope.setOutputSoapObject(rpc);
26.                  HttpTransportSE transportSE=new HttpTransportSE("http://192.168.0.2:9002/Test.asmx?wsdl");
27.                  try {
28.                      transportSE.call("http://tempuri.org/SaveInfo",envelope);
29.                      SoapObject soapObject=(SoapObject) envelope.bodyIn;
30.                      String result=soapObject.getProperty(0).toString();
31.                      Log.i("result", result);
32.                  } catch (IOException e) {
33.                      e.printStackTrace();
34.                  } catch (XmlPullParserException e) {
35.                      e.printStackTrace();
36.                  }
37.              };
38.          } .start();
39.      }
40.
41.      public static void GetInfo(final Callback callback) {
42.          new Thread() {
43.              @Override
44.              public void run() {
45.                  SoapSerializationEnvelope envelope=new SoapSerializationEnvelope(SoapEnvelope.VER10);
46.                  SoapObject rpc=new SoapObject("http://tempuri.org/","GetInfo");
47.                  envelope.bodyOut=rpc;
48.                  envelope.dotNet=true;
49.                  envelope.setOutputSoapObject(rpc);
50.                  HttpTransportSE transportSE=new HttpTransportSE("http://192.168.0.2:9002/Test.asmx?wsdl");
51.                  try {
52.                      transportSE.call("http://tempuri.org/GetInfo",envelope);
53.                      SoapObject object=new SoapObject();
54.                      object=(SoapObject) envelope.bodyIn;
55.                      String result=object.getProperty(0).toString();
56.                      Log.i("result", result);
57.                      callback.call(result);
```

```
58.                    } catch(IOException e) {
59.                        e.printStackTrace();
60.                    } catch(XmlPullParserException e) {
61.                        e.printStackTrace();
62.                    }
63.                }
64.            } .start();
65.        }
66. }
```

第 20 行创建 SoapSerializationEnvelope 对象时需要通过 SoapSerializationEnvelope 类的构造方法设置 SOAP 的版本号。该版本号需要根据服务器端 WebService 的版本号设置。

第 21 行实例化 SoapObject 对象，指定 WebService 的命名空间，以及调用方法名称。

第 22 行设置调用方法的参数值，这一步是可选的。如果方法没有参数，可以省略这一步。

第 23 行注册 envelope，表示提交请求信息，传出 SOAP 消息体。

第 24 行用来设置是否调用.NET 开发的 WebService。如果服务器用.NET 程序，就为 true。

第 25 行用来设置 SOAP 请求信息 envelope 发出的信息格式为 SoapObject 对象。

第 26 行构建传输对象，并指明 WSDL 文档的 URL。

第 28 行调用 WebService，其中第一个参数为 "nameSpace+方法名称"，第二个参数为 envelope 对象。

第 29、30 行解析返回数据。

第 57 行使用回调函数获得结果数据。

38　更新定时任务

11.3　更新活动 MainActivity 中的定时器任务

由于希望能够实时监测人体数据，因此需要更新活动 MainActivity 中的定时器任务，保证数据随时更新，添加如下代码。

```
1. if(bodyValue!=null) {
2.     if(bodyValue.equals("0")) {
3.         bodyValue=getResources().getString(R.string.breaking_abnormal);
4.         WebServiceHelper.SaveInfo(dateFormat.format(new Date())+" "+bodyValue);
5.     } else {
6.         bodyValue=getResources().getString(R.string.breaking_normal);
7.     }
8. }
```

从云平台获取人体数据，判断是否有人闯入，有人闯入则将报警信息通过 WebService 保存到远程数据库中。

第 1 行判断云平台传来的人体变量 bodyValue 的值是否不为空。

第 2、3 行判断 bodyValue 的值为 0 时，则将字符资源中的变量 breaking_abnormal 赋值给 bodyValue，breaking_abnormal 对应的中文为"有人闯入"。

第 4 行调用 WebService 的 SaveInfo()方法将"当前时间+bodyValue"的数据存放到 WebService 上。

第 6 行表示当云平台获取的 bodyValue 的值不为 0 时，将"正常"的字符串赋值给 bodyValue。

39　查看历史报警信息

11.4　查看历史报警信息

单击主界面中"禁入区域监控"区域下的警报信息，可以跳转到警报数据界面。警报数据界面如图 11-7 所示。该界面使用 ListView（列表框）组件进行显示。每一行显示的信息为人体红外传感器检测到有人的时间+"有人闯入"的提示信息。

图 11-7　警报数据界面

ListView 组件即列表视图，该组件以垂直列表的形式显示信息。创建该组件的语法如下。

```
1. <ListView
2.    //属性列表>
3. </ListView>
```

ListView 组件常见的 XML 属性如表 11-1 所示。

表 11-1 ListView 组件属性

XML 属性	说明
android:divider	设置分割条，可以用颜色或 drawable 资源分割
android:dividerHeight	设置分割条的高度
android:footerDividersEnabled	设置是否在 footer View 之前绘制分割条，默认值为 true，false 表示不绘制分割条
android:headerDividersEnabled	设置是否在 footer View 之后绘制分割条，默认值为 true，false 表示不绘制分割条
android:entries	通过数组资源为 ListView 指定列表项

和任务 8 中一样，这里我们使用自定义适配器为 ListView 绑定数据源，步骤如下。

1）新建 warn_list_item 布局文件，该布局文件为线性布局文件。

2）新建 activity_warnlist 布局文件，该布局文件包含 ListView 组件。

3）新建 WarnAdapter 自定义适配器。

4）新建 WarnListActivity 活动，该活动关联 activity_warnlist 布局文件。在该活动中实例化自定义适配器 WarnAdapter，并将适配器对象绑定到 ListView 组件上。

11.4.1 为 ListView 创建布局

新建布局文件之前，需要定义将用到的背景资源。在 drawable 目录下新建 shape.xml 与 shape_listview_background.xml 样式文件。shape.xml 用于指定 activity_warnlist 布局文件的背景，shape_listview_background.xml 用于指定 warn_list_item 布局文件的背景。

（1）样式文件 shape.xml 的代码如下。

```
1.  <?xml version="1.0" encoding="utf-8"?>
2.  <shape xmlns:android="http://schemas.android.com/apk/res/android">
3.      <gradient android:startColor="#F0F0F0"
4.          android:endColor="#F0F0F0"
5.          android:angle="90"/>
6.      <stroke
7.          android:width="2dp"
8.          android:color="#6C6C6C"/>
9.      <corners
10.         android:radius="10dip"/>
11.     <padding
12.         android:left="0dp"
13.         android:top="0dp"
14.         android:right="0dp"
15.         android:bottom="0dp"/>
16. </shape>
```

第 3~5 行中，gradient 表示颜色渐变，可设置从什么颜色渐变到什么颜色，渐变开始的角度、类型等。startColor 属性为渐变开始的颜色，endColor 为渐变结束的颜色，angle 为渐变角度，必须为 45 的整数倍。渐变默认的模式为 android:type="linear"，即线性渐变。如果指定渐变模式为径向渐变，则设置 android:type="radial"，径向渐变需要指定半径，如 android:gradientRadius="50"。

第 6～8 行中，stroke 可以理解为描边，width 属性用来设置边的宽度，color 属性用来设置边的颜色。

第 9、10 行中，corners 表示圆角，android:radius 为半径，值越大角越圆。

第 11～15 行中，padding 表示该属性所在的主组件中内部布局（子组件）的边距。

（2）样式文件 shape_listview_background.xml 的代码如下。

```xml
1.  <?xml version="1.0" encoding="utf-8"?>
2.  <shape xmlns:android="http://schemas.android.com/apk/res/android"
3.      android:shape="rectangle">
4.      <solid android:color="#00000000"/>
5.      <stroke
6.          android:width="1px"
7.          android:color="#00000000"/>
8.      <corners
9.          android:bottomLeftRadius="0dp"
10.         android:bottomRightRadius="0dp"
11.         android:topLeftRadius="0dp"
12.         android:topRightRadius="0dp"/>
13. </shape>
```

第 4 行中，solid 表示填充，只有一个属性即 color，表示填充的颜色。

在 drawable 文件夹中完成上述两个样式文件的创建后，在 Layout 文件夹中新建 warn_list_item.xml 布局文件，代码如下。

```xml
1.  <LinearLayout xmlns:android="http://schemas.android.com/apk/res/android"
2.      android:layout_width="match_parent"
3.      android:layout_height="wrap_content"
4.      android:gravity="center_vertical"
5.      android:padding="5dp"
6.      android:background="@drawable/shape_listview_background"
7.      android:orientation="horizontal">
8.      <TextView
9.          android:id="@+id/tv_item"
10.         android:textSize="18sp"
11.         android:layout_width="wrap_content"
12.         android:layout_height="wrap_content"
13.         android:text=" "
14.         />
15. </LinearLayout>
```

第 4 行中，android:gravity="center_vertical"表示当前线性布局中的元素垂直居中显示。android:gravity 是针对组件里的元素来说的，用来控制元素在该组件里的显示位置；android:layout_gravity 是针对组件本身而言，用来控制该组件在包含该组件的父组件中的位置。

第 5 行设置内部元素到边框的距离为 5dip。

第 6 行设置布局文件的背景。

第 7 行设置布局中的组件水平排列。

第 8~14 行在布局中添加一个文本组件。

上述布局文件完成后，在 Layout 文件夹中新建 activity_warnlist.xml 布局文件。activity_warnlist.xml 布局文件代码如下：

```xml
1. <?xml version="1.0" encoding="utf-8"?>
2. <LinearLayout
3.     xmlns:android="http://schemas.android.com/apk/res/android" android:layout_width="match_parent"
4.     android:layout_height="match_parent">
5.     <ListView
6.         android:id="@+id/list_warn"
7.         android:layout_width="match_parent"
8.         android:layout_height="wrap_content"
9.         android:background="@drawable/shape"
10.        android:cacheColorHint="#00000000"
11.        android:drawSelectorOnTop="false"
12.        android:fadingEdge="none"
13.        android:listSelector="#00000000"
14.        android:layout_marginLeft="10dip"
15.        android:layout_marginRight="10dip">
16.    </ListView>
17. </LinearLayout>
```

第 5~16 行创建一个 ListView 组件。

第 9 行为 ListView 组件设置背景，该背景是之前定义的样式文件 shape.xml。

第 10 行表示去除 ListView 组件的拖动背景色。ListView 组件在拖动的时候背景图片消失变成黑色背景，拖动完毕背景图片才显示出来，此时在 XML 中加入 android:cacheColorHint="#00000000"可解决该问题。

第 11 行表示按住某条记录不放，颜色会在记录的后面成为背景色，但是记录内容的文字是可见的。

第 12 行用来设置拖动滚动条时边框渐变的方向。值为 none 表示边框颜色不变，horizontal 表示水平方向颜色变淡，vertical 表示垂直方向颜色变淡。

第 13 行设置背景色，当按住 ListView 组件中的某个 item 时，会显示背景色（Android 系统默认设置的颜色），若想设置按住时无色（透明色，不用系统背景色），并设置自己的按住效果，则需添加 android:listSelector="#00000000"。

第 14、15 行设置组件距左右边框的边距均为 10dip。

11.4.2 创建自定义适配器 WarnAdapter

新建类 WarnAdapter 继承自 BaseAdapter。BaseAdapter 是一个抽象类，所以需要实现其中的 4 个抽象方法，分别是 getCount()、getItem()、getItemId()、getView()。WarnAdapter 类中还需实现自己的构造方法，其构造方法的参数为上下文对象、数据源。

```
1. package cn.edu.jsit.smartfactory.tools;
2.
```

```java
3.  import android.content.Context;
4.  import android.view.LayoutInflater;
5.  import android.view.View;
6.  import android.view.ViewGroup;
7.  import android.widget.BaseAdapter;
8.  import android.widget.TextView;
9.  import java.util.List;
10. import cn.edu.jsit.smartfactory.R;
11.
12. public class WarnAdapter extends BaseAdapter {
13.     private List<String> mDatas;
14.     private LayoutInflater mlayoutInflater;
15.
16.     public WarnAdapter(Context context, List<String> datas) {
17.         this.mDatas=datas;
18.         mlayoutInflater=LayoutInflater.from(context);
19.     }
20.
21.     @Override
22.     public int getCount() {
23.         return mDatas.size();
24.     }
25.
26.     @Override
27.     public Object getItem(int i) {
28.         return mDatas.get(i);
29.     }
30.
31.     @Override
32.     public long getItemId(int i) {
33.         return i;
34.     }
35.
36.     @Override
37.     public View getView(int i, View view, ViewGroup viewGroup) {
38.         ViewHolder holder=null;
39.         if(view==null) {
40.             view=mlayoutInflater.inflate(R.layout.warn_list_item, viewGroup, false);
41.             holder=new ViewHolder();
42.             holder.textView=view.findViewById(R.id.tv_item);
43.             view.setTag(holder);
44.         } else {
45.             holder=(ViewHolder) view.getTag();
46.         }
47.         holder.textView.setText(mDatas.get(i));
```

```
48.             return view;
49.         }
50.
51.     private class ViewHolder {
52.         TextView textView;
53.     }
54. }
```

第 12 行新建类 WarnAdapter 继承 BaseAdapter。

第 16~19 行定义了类 WarnAdapter 的构造函数。函数中第一个参数为上下文对象，第二个参数为数据源。

第 22~24 行中，getCount()方法的作用是返回 ListView 组件中 items 的总数。第 23 行返回了数据源的长度。

第 27~29 行中，getItem(int i)方法的作用是按照位置获取数据对象。参数 i 表示 item 的索引。

第 32、33 行中，getItemId(int i)方法的作用是根据位置获取 item 的索引。参数 i 表示 item 的索引。

第 37~49 行中，getView(int i, View view, ViewGroup viewGroup)方法的作用是返回每个 item 的显示效果。参数 i 为 item 的索引。参数 view 是指定位置要显示的视图，缓存池中有对应的缓存，viewGroup 参数在加载 XML 视图时使用。

第 51~53 行定义一个持有者的类，类中定义了一个文本组件。

第 38 行声明持有类 ViewHolder 对象 holder，ViewHolder 可将需要缓存的 view 封装好。

第 39 行表示如果 view 未被实例化过，缓存池中没有对应的缓存。

第 40 行从 warn_list_item.xml 布局文件中反射组件并赋值给 view。此处调用的 LayoutInflater 类中的 inflate(int resource, ViewGroup root, boolean attachToRoot)方法，该方法共有 3 个参数。第一个参数 resource，指的是需要加载布局资源文件的 id，即需要将这个布局文件加载到 Activity 中来操作。第二个参数 root，指的是需要附加到 resource 资源文件的根组件，即 inflate()方法会返回一个 View 对象，如果第三个参数 attachToRoot 为 true，就将这个 root 作为根对象返回，否则仅将这个 root 对象的 LayoutParams 宽高参数属性附加到 resource 对象的根布局对象上，也就是布局文件 resource 的最外层的 View 上，如一个 LinearLayout 或者其他的 Layout 对象。如果没有根组件，就写 null。第三个参数 attachToRoot，指的是是否将 root 附加到布局文件的根视图上，取值可以为 true 或 false。

第 41 行实例化持有类对象 holder。

第 42 行中，view 中返回组件 id 为 tv_item 的文本组件，并与持有类中的文本框组件 textView 相关联。

第 43 行中，convertView 的 setTag()方法将这些 view 缓存起来供下次调用。

第 44~46 行中，当实例化的 view 不为空时，ViewHolder 被复用。

第 47 行为 warn_list_item.xml 布局文件中的文本组件赋值，值为数据源 mDatas 中的值。

第 48 行返回实例化后的 view。

11.4.3 创建活动 WarnListActivity

```
1. package cn.edu.jsit.smartfactory;
2.
3. import android.app.Activity;
4. import android.os.Bundle;
5. import android.os.Handler;
6. import android.os.Message;
7. import android.support.annotation.Nullable;
8. import android.widget.ListView;
9. import org.json.JSONArray;
10. import org.json.JSONException;
11. import org.json.JSONObject;
12. import java.util.ArrayList;
13. import java.util.List;
14. import cn.edu.jsit.smartfactory.tools.WarnAdapter;
15. import cn.edu.jsit.smartfactory.tools.WebServiceHelper;
16.
17. public class WarnListActivity extends Activity {
18.     private ListView listView;
19.     private List<String> msgs=new ArrayList<>();
20.     private Handler handler=new Handler() {
21.         @Override
22.         public void handleMessage(Message msg) {
23.             if(msg.what==0) {
24.                 WarnAdapter adapter=new WarnAdapter(WarnListActivity.this, msgs);
25.                 listView.setAdapter(adapter);
26.             }
27.         }
28.     };
29.
30.     @Override
31.     protected void onCreate(@Nullable Bundle savedInstanceState) {
32.         super.onCreate(savedInstanceState);
33.         setContentView(R.layout.activity_warnlist);
34.         listView=findViewById(R.id.list_warn);
35.         WebServiceHelper.GetInfo(new WebServiceHelper.Callback() {
36.             @Override
37.             public void call(String s) {
38.                 initWeb(s);
39.             }
40.         });
41.     }
42.
43.     protected void initWeb(String result) {
```

```
44.         try {
45.             JSONArray array=new JSONArray(result);
46.             for(int i=0; i<array.length(); i++) {
47.                 JSONObject object=(JSONObject) array.get(i);
48.                 String msg=object.getString("Message");
49.                 msgs.add(msg);
50.             }
51.             handler.sendEmptyMessage(0);
52.         } catch(JSONException e) {
53.             e.printStackTrace();
54.         }
55.     }
56. }
```

第 17 行新建一个类 WarnListActivity 继承 Activity。

第 19 行实例化一个 String 类型的数组 msgs。

第 20~28 行使用 Handler 机制，判断当收到的消息 msg 的 what 属性为 0 时，实例化自定义的适配器对象，并绑定 ListView 组件，从而实现 ListView 组件中视图的显示。

第 33 行指定当前 Activity 启动 activity_warnlist 布局。

第 35~40 行调用回调函数实现服务器返回结果的 JSON 解析。

第 43~55 行定义 initWeb()方法，实现服务器返回结果的 JSON 解析。

在 activity_main.xml 布局文件中，为报警信息 TextView 添加 onClick 属性的代码如下。

```
1. <TextView
2.     android:id="@+id/tv_breaking_value"
3.     ...
4.     android:onClick="onClick"
5. />
```

在 MainActivity 中更新 onClick()方法的代码如下。

```
1.  @Override
2.  public void onClick(View view){
3.      Intent intent = new Intent(MainActivity.this, DataChartActivity.class);
4.      switch(view.getId()){
5.          ...
6.          case R.id.tv_breaking_value:
7.              startActivity(new Intent(MainActivity.this,WarnListActivity.class));
8.              break;
9.      }
10. }
```

任务 12　创建摄像头监控界面

任务概述

单击"开启监控"按钮，连接摄像头，此时按钮显示内容变为"关闭监控"。单击开启/关闭监控按钮右边的上、下、左、右箭头按钮，可以实现控制摄像头的上、下、左、右转动，如图 12-1 所示。在摄像头监控界面中通过 WebView 组件显示摄像头画面。

图 12-1　摄像头监控界面

知识目标

- 掌握 WebView 组件。
- 掌握 HttpURLConnection 类。

技能目标

- 能使用 HTTP 访问和控制摄像头。
- 能使用 include 标签重用布局。

40　创建摄像头监控布局文件

12.1　创建摄像头监控布局文件

我们使用 HTTP 的 URL 地址访问网络摄像头的画面，因此需要用 WebView 组件来显示网络摄像头返回的画面。

WebView（网络视图）组件是一个基于 webkit 引擎、展现 Web 页面的组件。WebView 组件功能强大，除了具有一般 View 的属性和设置外，还可以对 URL 请求、页面加载、渲染、页面交互进行强大的处理，Android 4.4 版本之后，直接使用 Chrome 作为内置网页浏览器。

WebView 组件常用的方法如表 12-1 所示。

表 12-1 WebView 组件常用方法

方法	说明
loadUrl(String url)	加载 URL 信息，URL 可以是网络地址，也可以是本地网络文件
goBack()	向后浏览历史页面
goForward()	向前浏览历史页面
loadData(String data,String mimeType, String encoding)	用于将指定的字符串数据加载到浏览器中
reload()	用于刷新当前页面

使用 WebView 组件之前，不要忘记在清单文件中声明访问网络的权限。

```
<uses-permission android:name="android.permission.INTERNET"/>
```

12.1.1 创建摄像头布局文件

了解 WebView 组件后，我们现在来使用它。在 layout 目录下新建摄像头布局文件 activity_camera.xml，代码如下。

```
1.  <?xml version="1.0" encoding="utf-8"?>
2.  <LinearLayout
3.      xmlns:android="http://schemas.android.com/apk/res/android"
4.      android:orientation="vertical"
5.      android:layout_width="match_parent"
6.      android:layout_height="match_parent"
7.      android:gravity="center_horizontal">
8.      <WebView
9.          android:id="@+id/webView"
10.         android:layout_width="match_parent"
11.         android:layout_height="0dp"
12.         android:layout_weight="2"
13.         >
14.     </WebView>
15.     <LinearLayout
16.         android:layout_width="match_parent"
17.         android:layout_height="0dp"
18.         android:layout_weight="1"
19.         >
20.         <LinearLayout
21.             android:layout_width="0dp"
```

```
22.            android:layout_height="match_parent"
23.            android:layout_weight="1"
24.            android:gravity="center"
25.            >
26.            <Button
27.                android:id="@+id/btn_camera"
28.                android:layout_width="match_parent"
29.                android:layout_margin="20dp"
30.                android:layout_height="wrap_content"
31.                android:text="@string/start_camera"
32.                android:background="@drawable/bg_login_blue"
33.                android:textSize="20sp"
34.                android:textColor="@color/colorWhite"
35.                />
36.        </LinearLayout>
37.        <include
38.            android:layout_weight="1"
39.            android:layout_height="match_parent"
40.            android:layout_width="0dp"
41.            layout="@layout/view_ptz_direction"
42.            />
43.    </LinearLayout>
44. </LinearLayout>
```

第 4 行设置当前线性布局中各布局/组件的排列方式为垂直排列。

第 7 行设置当前线性布局中的内容水平居中放置。

第 8~14 行创建一个 WebView 组件,并且该组件的高度占整个父容器(最外层线性布局)高度的 2 份。

第 15~19 行创建一个线性布局,该布局的高度占整个父容器(最外层线性布局)高度的 1 份。

第 20~36 行在前面线性布局中创建一个子线性布局,在该线性布局中主要放置的是开启监控的按钮。

第 32 行设置 Button 的背景,背景的资源文件为 bg_login_blue.xml,该文件的代码后面会介绍。

第 37~42 行使用<include>标签来重用布局代码,其中布局文件 view_ptz_direction.xml 为控制摄像头上、下、左、右转动的布局,如图 12-2 所示。

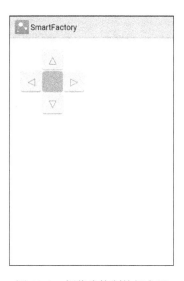

图 12-2 摄像头控制按钮布局

<include>标签可以实现在一个布局中引用另一个布局,这通常适合于界面布局复杂、不同界面有共用布局的 APP 中,如一个 APP 的顶部布局、侧边栏布局、底部 Tab 栏布局、ListView 和 GridView 每一项的布局等,将这些同一个 APP 中有多个界面用到的布局抽取出来再通过<include>标签引用,既可以降低 layout 的复杂度,又可以做到布局重用。<include>标签的使用很简单,只要在布局文件中

需要引用其他布局的地方，使用 layout= "@layout/child_layout"即可实现。

12.1.2 创建摄像头控制按钮布局文件

创建 view_ptz_direction.xml 布局文件控制摄像头的旋转方向，代码如下。

```
1.  <?xml version="1.0" encoding="utf-8"?>
2.  <RelativeLayout xmlns:android="http://schemas.android.com/apk/res/android"
3.      android:layout_width="150dp"
4.      android:layout_height="150dp"
5.      android:layout_alignParentBottom="true"
6.      android:layout_margin="25dp">
7.
8.      <TextView
9.          android:id="@+id/btn_ptz_up"
10.         android:layout_width="50dp"
11.         android:layout_height="50dp"
12.         android:gravity="center"
13.         android:layout_above="@+id/btn_handle_center"
14.         android:layout_centerHorizontal="true"
15.         android:background="@drawable/bg_btn_ptz"
16.         android:textColor="@color/text_blue"
17.         android:textSize="30sp"
18.         android:text="△"/>
19.
20.     <TextView
21.         android:id="@+id/btn_ptz_left"
22.         android:layout_width="50dp"
23.         android:layout_height="50dp"
24.         android:gravity="center"
25.         android:layout_toLeftOf="@+id/btn_handle_center"
26.         android:layout_centerVertical="true"
27.         android:background="@drawable/bg_btn_ptz"
28.         android:textColor="@color/text_blue"
29.         android:textSize="30sp"
30.         android:text="◁ "/>
31.
32.     <TextView
33.         android:id="@+id/btn_ptz_right"
34.         android:layout_width="50dp"
35.         android:layout_height="50dp"
36.         android:gravity="center"
37.         android:layout_toRightOf="@+id/btn_handle_center"
38.         android:layout_centerVertical="true"
39.         android:background="@drawable/bg_btn_ptz"
40.         android:textColor="@color/text_blue"
41.         android:textSize="30sp"
```

```
42.        android:text="▷ "/>
43.
44.    <TextView
45.        android:id="@+id/btn_ptz_down"
46.        android:layout_width="50dp"
47.        android:layout_height="50dp"
48.        android:gravity="center"
49.        android:layout_below="@+id/btn_handle_center"
50.        android:layout_centerHorizontal="true"
51.        android:background="@drawable/bg_btn_ptz"
52.        android:textColor="@color/text_blue"
53.        android:textSize="30sp"
54.        android:text="▽ "/>
55.
56.    <TextView
57.        android:id="@+id/btn_handle_center"
58.        android:layout_width="50dp"
59.        android:layout_height="50dp"
60.        android:layout_centerInParent="true"
61.        android:background="@drawable/bg_grid_transparent"/>
62.
63. </RelativeLayout>
```

第 16、28、40、52 行为当前的 TextView 组件设置背景，背景文件为 bg_btn_ptz.xml。

第 61 行设置图 12-2 中控制按钮中间的 TextView 组件的背景，背景文件为 bg_grid_transparent.xml。

上面提到背景文件需要单独创建。

1）在 drawable 下新建 bg_btn_ptz.xml 文件，代码如下。

```
1. <?xml version="1.0" encoding="utf-8"?>
2. <selector xmlns:android="http://schemas.android.com/apk/res/android">
3.     <item
4.         android:state_focused="false"
5.         android:drawable="@drawable/generate_btn"/>
6.     <item
7.         android:state_focused="true"
8.         android:drawable="@drawable/generate_btn_sel"/>
9.     <item
10.        android:state_selected="true"
11.        android:drawable="@drawable/generate_btn_sel"/>
12.    <item
13.        android:drawable="@drawable/generate_btn"/>
14. </selector>
```

bg_btn_ptz.xml 文件中主要定义了一个 selector 状态选择器，在 Android 开发中常常使用 selector 来设置组件的背景。这样就可不使用代码来控制组件在不同状态下不同背景或图片的变化，使用非常方便。selector 是在 drawable/xxx.xml 中配置的。selector 又称状态列表，分为两种，分别是 Color-selector 和 Drawable-selector。此处我们使用 Drawable-selector，Drawable-selector 是背景图状态列表，背景会根据组件的状态变化而变化。具体实施之前需先在 drawable 目录下导入本任务资源包中的图片 generate_btn、generate_btn_sel。常用的状态设置类型具体如表 12-2 所示。

表 12-2 状态设置类型

属性	说明
android:state_pressed	设置是否为按压状态，一般在 true 时设置该属性，表示已按压状态，默认为 false
android:state_selected	设置是否为选中状态，true 表示已选中，false 表示未选中
android:state_checked	设置是否为勾选状态，主要用于 CheckBox 和 RadioButton，true 表示已被勾选，false 表示未被勾选
android:state_checkable	设置勾选是否为可用状态，类似 state_enabled，state_enabled 会影响触屏或单击事件，state_checkable 影响勾选事件
android:state_focused	设置是否为获得焦点状态，true 表示获得焦点，默认为 false，表示未获得焦点
android:state_enabled	设置触屏或单击事件是否为可用状态，一般只在 false 时设置该属性，表示不可用状态

2）在 drawable 下新建 bg_grid_transparent.xml 文件，代码如下。

```
1.  <?xml version="1.0" encoding="utf-8"?>
2.  <shape xmlns:android="http://schemas.android.com/apk/res/android">
3.      <stroke
4.          android:width="1dp"
5.          android:color="#edeeee"/>
6.      <corners
7.          android:radius="10dp"
8.          android:topLeftRadius="6dp"
9.          android:topRightRadius="6dp"
10.         android:bottomLeftRadius="6dp"
11.         android:bottomRightRadius="6dp"/><!-- 设置圆角半径 -->
12.     <solid android:color="#50000000"/>
13. </shape>
```

上述文件中主要定义了<shape>标签，该标签用来定义形状的样式，共有 6 个属性，每个属性的具体说明如表 12-3 所示。

表 12-3 <shape>标签属性

属性	说明
corners	设置圆角，即四个角的弧度
gradient	设置颜色渐变
padding	定义内容离边界的距离
size	设置形状的长宽
solid	设置填充颜色
stroke	设置图片边缘颜色

我们同样需要使用 selector 与 shape 给按钮设置样式。在 drawable 目录下新建三个文件，分别为 bg_login_blue.xml、bg_btn_bluecorners.xml、bg_btn_bluecorners_press.xml。

1) 创建 bg_login_blue.xml 文件，代码如下。

```
1.  <?xml version="1.0" encoding="utf-8"?>
2.  <selector xmlns:android="http://schemas.android.com/apk/res/android">
3.      <item
4.          android:state_pressed="false"
5.          android:drawable="@drawable/bg_btn_bluecorners"/>
6.      <item
7.          android:state_pressed="true"
8.          android:drawable="@drawable/bg_btn_bluecorners_press"/>
9.      <item
10.         android:drawable="@drawable/bg_btn_bluecorners"/>
11. </selector>
```

第 4 行设置触屏状态为不触屏。

第 5 行导入 drawable 文件夹中的 bg_btn_bluecorners.xml 作为不触屏时组件的背景样式。

第 7 行设置触屏状态为触屏。

第 8 行导入 drawable 文件夹中的 bg_btn_bluecorners_press.xml 作为触屏时组件的背景样式。

2) 创建 bg_btn_bluecorners.xml，代码如下。

```
1.  <?xml version="1.0" encoding="utf-8"?>
2.  <shape xmlns:android="http://schemas.android.com/apk/res/android">
3.      <corners
4.          android:radius="9dp"
5.          android:topLeftRadius="12dp"
6.          android:topRightRadius="12dp"
7.          android:bottomLeftRadius="12dp"
8.          android:bottomRightRadius="12dp"/><!-- 设置圆角半径 -->
9.      <solid android:color="@color/text_blue"/>
10. </shape>
```

第 4 行设置圆角的半径，值越大角越圆。

第 5~8 行设置圆角左上、右上、左下、右下的半径。

第 9 行设置填充的颜色为 strings.xml 资源文件所定义的颜色。

3) 创建 bg_btn_bluecorners_press.xml，代码如下。

```
1.  <?xml version="1.0" encoding="utf-8"?>
2.  <shape xmlns:android="http://schemas.android.com/apk/res/android">
3.      <corners
4.          android:radius="9dp"
5.          android:topLeftRadius="12dp"
```

```
6.        android:topRightRadius="12dp"
7.        android:bottomLeftRadius="12dp"
8.        android:bottomRightRadius="12dp"/><!-- 设置圆角半径 -->
9.    <solid android:color="#2a65a6"/>
10. </shape>
```

第 4 行设置圆角的半径，值越大角越圆。

第 5~8 行设置圆角左上、右上、左下、右下的半径。

第 9 行设置填充的颜色。

41 创建 HTTP 访问类 HttpRequest

12.2 创建 HTTP 访问类 HttpRequest

HTTP 是 Hyper Text Transfer Protocol（超文本传输协议）的缩写，是用于从网络服务器传输超文本到本地浏览器的传送协议。HTTP 属于应用层的面向对象的协议，由于其简捷、快速的方式，适用于分布式超媒体信息系统。

HTTP 是基于 TCP/IP 的应用层协议。TCP/IP 是传输层协议，主要解决数据如何在网络中传输，HTTP 是应用层协议，主要解决如何包装数据。

HTTP 的工作流程如下。

1）用户单击浏览器上的 URL（超链接），Web 浏览器与 Web 服务器建立连接。

2）建立连接后，客户端发送请求给服务器端，请求的格式为：统一资源标识符（URL）+协议版本号（一般是 1.1）+MIME 信息（多个消息头）+一个空行。

3）服务器端收到请求后，给予相应的返回信息，返回格式为：协议版本号+状态行（处理结果）+多个信息头+空行+实体内容（如返回的 HTML）。

4）客户端接收服务器端返回的信息，通过浏览器显示出来，然后与服务器端断开连接。当然如果中途某步发生错误的话，错误信息会返回到客户端并显示，如经典的 404 错误信息。

如果读者对上面的流程还不清晰，可以使用 HttpWatch 或者 Firefox 抓包查看请求和响应的信息。HTTP 请求包含的内容如图 12-3 所示。

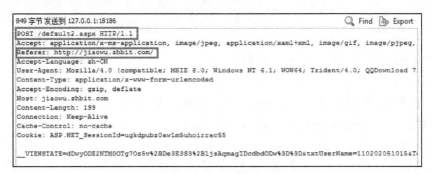

图 12-3 HTTP 请求内容

HTTP 响应包含的内容如图 12-4 所示。

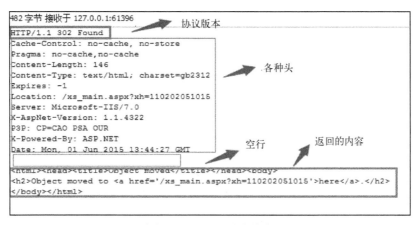

图 12-4 HTTP 响应内容

HTTP 的业务流程如图 12-5 所示。

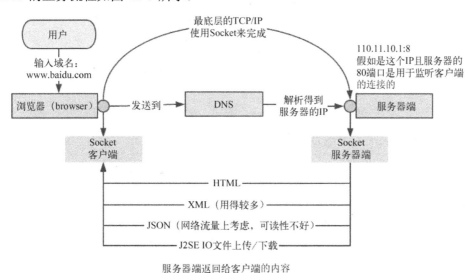

图 12-5 HTTP 业务流程

Android 系统提供了两种 HTTP 通信类：HttpURLConnection 和 HttpClient。两者都支持 HTTPS、流式传输上传和下载、可配置超时、IPv6 和连接池。不过由于 HttpClient 存在的 API 数量过多、扩展困难等缺陷，在开发中也不会推荐使用这种方式。因此在 Android M（6.0 版本）系统中，HttpClient 的功能被完全移除了，标志着此功能被正式弃用，所以在此介绍官方建议的 HttpURLConnection 的用法。

创建 HTTP 访问类 HttpRequest.java，代码如下。

```
1. package cn.edu.jsit.smartfactory.tools;
2. import java.io.IOException;
3. import java.net.HttpURLConnection;
4. import java.net.MalformedURLException;
5. import java.net.URL;
6. public class HttpRequest {
```

```java
7.      public synchronized static void send(final String s) {
8.          new Thread() {
9.              public void run() {
10.                 try {
11.                     URL url=new URL(s);
12.                     HttpURLConnection connection=(HttpURLConnection) url.openConnection();
13.                     connection.connect();
14.                     int size=0;
15.                     while(size==0) {
16.                         size=connection.getInputStream().available();
17.                     }
18.                     byte[] msg=new byte[size];
19.                     connection.getInputStream().read(msg);
20.                 } catch (MalformedURLException e) {
21.                     //TODO Auto-generated catch block
22.                     e.printStackTrace();
23.                 } catch (IOException e) {
24.                     //TODO Auto-generated catch block
25.                     e.printStackTrace();
26.                 }
27.             };
28.         } .start();
29.     }
30. }
```

第 7 行中，静态方法 send() 被 synchronized 同步锁修饰。synchronized 用于多线程设计，有了 synchronized 关键字，多线程程序的运行结果将变得可以控制。synchronized 关键字用于保护共享数据。synchronized 依靠"锁"机制进行多线程同步，"锁"有两种，一种是对象锁，一种是类锁。

Synchronized 可修饰普通方法和静态方法。其具体区别如下。

- ◆ 当修饰普通方法时，synchronized 是对类的当前实例（当前对象）进行加锁，防止其他线程同时访问该类的该实例的所有 synchronized 块，注意这里是"类的当前实例"，类的两个不同实例就没有这种约束了。

- ◆ 当修饰静态方法时，synchronized 恰好就是要控制类的所有实例的并发访问，static synchronized 是限制多线程中该类的所有实例同时访问 JVM 中该类所对应的代码块。实际上，在类中如果某方法或某代码块中有 synchronized，那么在生成一个该类实例后，该实例也就有一个监视块，防止线程并发访问该实例的 synchronized 保护块，而 static synchronized 则是所有该类的所有实例公用的一个监视块。

第 8～28 行创建一个线程并启动。

第 11 行创建远程 URL 连接对象。

第 12 行通过远程 URL 连接对象打开一个连接，强制转换成 httpURLConnection 类型。

第 13 行发送请求。

第 14～19 行通过 connection 连接，获取输入流。

12.3 实现摄像头访问

42　实现摄像头访问

1）在 res/menu 目录下的 menu_main.xml 菜单文件中添加 item，代码如下。

```
1. <menu xmlns:android="http://schemas.android.com/apk/res/android"
2. xmlns:tools="http://schemas.android.com/tools"
3. xmlns:app="http://schemas.android.com/apk/res-auto"
4. tools:context=".MainActivity">
5.     <item android:id="@+id/action_setting"
6.         android:title="@string/globle_params_setting"
7.         android:icon="@drawable/setting"
8.         android:orderInCategory="1"
9.         app:showAsAction="never"/>
10.    <item android:id="@+id/action_camera"
11.        android:title="@string/monitor_camera"
12.        android:icon="@drawable/setting"
13.        android:orderInCategory="2"
14.        app:showAsAction="never"/>
15. </menu>
```

2）在 MainActivity 中的弹出菜单中单击菜单项进行界面跳转，代码如下。

```
1. @Override
2. public boolean onOptionsItemSelected(MenuItem menuItem) {
3.     switch (menuItem.getItemId()) {
4.         ...
5.         case R.id.action_camera:
6.             startActivity(new Intent(MainActivity.this, MonitorActivity.class));
7.         ...
8.     }
9. }
```

3）新建 MonitorActivity.java 并在 AndroidManifest 文件中注册 Activity，代码如下。

```
1. package cn.edu.jsit.smartfactory;
2. import android.app.Activity;
3. import android.hardware.camera2.CameraManager;
4. import android.os.Bundle;
5. import android.os.Handler;
6. import android.os.Message;
7. import android.support.annotation.Nullable;
8. import android.view.View;
```

```
9.   import android.webkit.WebSettings;
10.  import android.webkit.WebView;
11.  import android.widget.Button;
12.  import android.widget.TextView;
13.  import cn.edu.jsit.smartfactory.tools.HttpRequest;
14.  import cn.edu.jsit.smartfactory.tools.SmartFactoryApplication;
15.  public class MonitorActivity extends Activity implements View.OnClickListener {
16.      private WebView webView;
17.      private TextView btn_up,btn_down,btn_left,btn_right;
18.      private Button toggle_camera;
19.      private SmartFactoryApplication smartFactory;
20.      private LoadThread thread;
21.      private Boolean camera_state=false;
22.      Handler mHandler=new Handler() {
23.          @Override
24.          public void handleMessage(Message msg) {
25.              if(msg.what==1)
26.                  if(camera_state) {
27.                      webView.loadUrl("http://"+smartFactory.getCameraAddress()
28.                          +"/snapshot.cgi?user=admin&pwd=&strm=0&resolution=32");
29.                  }
30.          }
31.      };
32.
33.      @Override
34.      protected void onCreate(@Nullable Bundle savedInstanceState) {
35.          super.onCreate(savedInstanceState);
36.          setContentView(R.layout.activity_camera);
37.          smartFactory=(SmartFactoryApplication) getApplication();
38.          initView();
39.          thread=new LoadThread();
40.          thread.start();
41.      }
42.
43.      private void initView() {
44.          btn_up=findViewById(R.id.btn_ptz_up);
45.          btn_down=findViewById(R.id.btn_ptz_down);
46.          btn_left=findViewById(R.id.btn_ptz_left);
47.          btn_right=findViewById(R.id.btn_ptz_right);
48.          toggle_camera=findViewById(R.id.btn_camera);
49.          btn_up.setOnClickListener(this);
50.          btn_down.setOnClickListener(this);
51.          btn_left.setOnClickListener(this);
52.          btn_right.setOnClickListener(this);
53.          toggle_camera.setOnClickListener(this);
```

```java
54.         webView=findViewById(R.id.webView);
55.         webView.getSettings().setJavaScriptEnabled(true);
56.         webView.getSettings().setCacheMode(WebSettings.LOAD_NO_CACHE);
57.     }
58.
59.     @Override
60.     public void onClick(View view) {
61.         switch(view.getId()) {
62.         case R.id.btn_camera:
63.             if(camera_state) {
64.                 camera_state=false;
65.                 toggle_camera.setText("打开监控");
66.             } else {
67.                 camera_state=true;
68.                 toggle_camera.setText("关闭监控");
69.             }
70.             break;
71.         case R.id.btn_ptz_up:
72.             HttpRequest.
73.             send("http://"+smartFactory.getCameraAddress()+
74.                 "/decoder_control.cgi?command=2&user=admin&pwd=");
75.             break;
76.         case R.id.btn_ptz_down:
77.             HttpRequest.
78.             send("http://"+smartFactory.getCameraAddress()+
79.                 "/decoder_control.cgi?command=0&user=admin&pwd=");
80.             break;
81.         case R.id.btn_ptz_left:
82.             HttpRequest.
83.             send("http://"+smartFactory.getCameraAddress()+
84.                 "/decoder_control.cgi?command=6&user=admin&pwd=");
85.             break;
86.         case R.id.btn_ptz_right:
87.             HttpRequest.
88.             send("http://"+smartFactory.getCameraAddress()+
89.                 "/decoder_control.cgi?command=4&user=admin&pwd=");
90.             break;
91.         }
92.     }
93.
94.     public class LoadThread extends Thread {
95.         @Override
96.         public void run() {
97.             //TODO Auto-generated method stub
98.             while (true) {
```

```
99.                    try {
100.                        Thread.sleep(500);//线程暂停 0.5s（500ms）
101.                        Message message=new Message();
102.                        message.what=1;
103.                        mHandler.sendMessage(message);//发送消息
104.                    } catch(InterruptedException e) {
105.                        //TODO Auto-generated catch block
106.                        e.printStackTrace();
107.                    }
108.                }
109.            }
110.        }
111.
112.        @Override
113.        protected void onDestroy() {
114.            camera_state=false;
115.            super.onDestroy();
116.        }
117.    }
```

第 13 行导入 cn.edu.jsit.smartfactory 包中 tools 文件夹中的 HttpRequest 类，该类用于 HTTP 请求。

第 14 行导入 cn.edu.jsit.smartfactory 包中 tools 文件夹中的 SmartFactoryApplication 类，该类为全局变量的设置。

第 15 行创建一个名为 MonitorActivity 的类，继承 Activity，并实现 View 的 OnClickListener 接口。

第 22~31 行使用 Handler 机制，判断当收到的消息 msg 的 what 属性值为 1 时，且表示摄像头状态的变量值为 true 时，加载 URL 信息。

第 37 行实例化 SmartFactoryApplication 类的对象为 smartFactory。

第 38 行调用初始化组件的函数 initView()。

第 39、40 行创建一个线程 thread，并启动该线程。

第 43~57 行为 initView()函数的创建。

第 44~47 行获取摄像头监控布局中控制摄像头上、下、左、右转动的文本组件。

第 48 行获取摄像头监控布局中开启监控的按钮组件。

第 49~53 行为界面中开启监控按钮以及上、下、左、右文本组件添加监控。

第 54 行获取 WebView 组件。

第 55 行设置 WebView 是否允许执行 JavaScript 脚本，默认为 false，表示不允许，此处设置为允许。

第 56 行设置缓存模式。WebView 会将用户浏览过的网页 URL 以及网页文件（css、图片、js 等）保存到数据库表中。WebView 中存在着两种缓存：网页数据缓存（存储打开过的页面及资源）和 H5 缓存（即 AppCache）。缓存的模式共有 5 种，具体如表 12-4 所示。此处缓存模式设置为 LOAD_NO_CACHE。

表 12-4 缓存模式

缓存模式	含义
LOAD_CACHE_ONLY	不使用网络，只读取本地缓存数据
LOAD_DEFAULT	根据 cache-control 决定是否从网络上获取数据
LOAD_CACHE_NORMAL	API level 17 中已经废弃，从 API level 11 开始其作用同 LOAD_DEFAULT 模式
LOAD_NO_CACHE	不使用缓存，只从网络获取数据
LOAD_CACHE_ELSE_NETWORK	只要本地有，无论是否过期，或者 no-cache，都使用缓存中的数据

第 59~92 行为各个组件添加监听器。

第 61 行获取被单击的组件的 id。

第 62~70 行判断如果 id 为 btn_camera，先判断表示摄像头状态的变量 camera_state 是否为 true，如果为 true，将 camera_state 变量设置为 false，更新 Button 组件 toggle_camera 的显示内容为"打开监控"。如果 camera_state 变量为 false，则将 camera_state 变量设置为 true，更新 Button 组件 toggle_camera 的显示内容为"关闭监控"。

第 71~75 行判断如果单击时获取的组件 id 为 btn_ptz_up，则调用 HttpRequest 类中的 send()方法。send()方法中传入的参数构成为"摄像头 IP 地址+控制摄像头向上转动以及登录用户名、密码的指令"，从而达到连接摄像头并控制摄像头向上转动的功能。

第 76~80 行判断如果单击时获取的组件 id 为 btn_ptz_down，则调用 HttpRequest 类中的 send()方法。send()方法中传入的参数构成为"摄像头 IP 地址+控制摄像头向下转动以及登录用户名、密码的指令"，从而达到连接摄像头并控制摄像头向下转动的功能。

第 81~85 行判断如果单击时获取的组件 id 为 btn_ptz_left，则调用 HttpRequest 类中的 send()方法。send()方法中传入的参数构成为"摄像头 IP 地址+控制摄像头向左转动以及登录用户名、密码的指令"，从而达到连接摄像头并控制摄像头向左转动的功能。

第 86~90 行判断如果单击时获取的组件 id 为 btn_ptz_right，则调用 HttpRequest 类中的 send()方法。send()方法中传入的参数构成为"摄像头 IP 地址+控制摄像头向右转动以及登录用户名、密码的指令"，从而达到连接摄像头并控制摄像头向右转动的功能。

第 94~110 行创建一个线程类 LoadThread，实现每 0.5s 发送一次消息。

第 112~116 行重写 onDestroy()方法，当销毁当前 Activity 时，设置变量 camera_state 为 false。

创建完 MonitorActivity 活动后要在 AndroidManifest.xml 文件中注册，代码如下。

```
1.  <?xml version="1.0" encoding="utf-8"?>
2.  <manifest xmlns:android="http://schemas.android.com/apk/res/android"
3.      package="cn.edu.jsit.smartfactory">
4.      <uses-permission android:name="android.permission.INTERNET"/>
5.      <application
6.          android:allowBackup="true"
7.          android:icon="@mipmap/icon_launcher"
8.          android:label="@string/app_name"
9.          android:supportsRtl="true"
10.         android:theme="@style/Theme.AppCompat.Light.NoActionBar"
```

```xml
11.         android:name=".tools.SmartFactoryApplication">
12.         <activity android:name=".SplashActivity"
13.             android:configChanges="orientation|screenSize">
14.             <intent-filter>
15.                 <action android:name="android.intent.action.MAIN"/>
16.                 <category android:name="android.intent.category.LAUNCHER"/>
17.             </intent-filter>
18.         </activity>
19.         <activity
20.             android:name=".MainActivity"
21.             android:label="@string/main_page"
22.             android:theme="@style/Theme.AppCompat.Light.NoActionBar"
23.             android:launchMode="singleTask"
24.             >
25.             <intent-filter>
26.                 <category android:name="android.intent.category.DEFAULT"/>
27.                 <action android:name="android.intent.action.VIEW"/>
28.                 <data android:scheme="push"/>
29.             </intent-filter>
30.         </activity>
31.         />
32.         <activity
33.             android:name=".SettingActivity"
34.             android:label="@string/globle_params_setting"/>
35.         <activity android:name=".DataChartActivity"
36.             android:label="@string/chart_data" />
37.         <activity android:name=".WarnListActivity"
38.             android:label="@string/warn_data" />
39.         <activity android:name=".MonitorActivity"
40.             android:label="@string/monitor_camera" />
41.         <activity android:name=".UserActivity"/>
42.         <activity android:name=".LanguageActivity"/>
43.         <activity android:name=".AboutActivity"/>
44.     </application>
45. </manifest>
```

第 39、40 行为 MonitorActivity 的注册代码。

任务 13　创建抽屉导航

任务概述

创建抽屉导航栏，实现对程序的管理，丰富程序的功能。单击主界面左上角的菜单图标，弹出抽屉导航界面。此界面中包含个人设置、语言选择、关于软件、切换账户和退出程序等菜单项，用户可以通过单击侧滑栏中的菜单项跳转到对应的功能界面，如图 13-1 和图 13-2 所示。

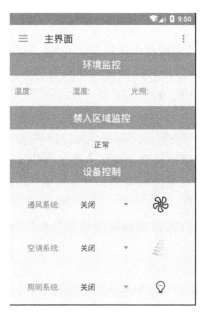

图 13-1　主界面　　　　　　　　　　　图 13-2　抽屉导航

知识目标
- 掌握 ToolBar 组件。
- 掌握 DrawLayout 组件。
- 掌握 NavigationView 组件。

技能目标
- 能使用 ToolBar、DrawLayout 和 NavigationView 组件实现抽屉导航。

43　创建抽屉导航

13.1　使用 ToolBar、DrawLayout 和 NavigationView 创建抽屉导航

我们使用 ToolBar+DrawerLayout+NavigationView 的组合实现上面侧滑抽屉的效果。Toolbar 是 Google 在 Android 5.0 中推出的 Material Design 风格的一款导航组件，以取代

179

Android 3.0 开始流行的 Actionbar（Actionbar 不能定制）。Toolbar 是应用内容的标准工具栏，可以说是 Actionbar 的升级版。相较于 Actionbar，Toolbar 最明显的优点就是变得很自由，可随处放置。Toolbar 组件常见的属性如表 13-1 所示。

表 13-1 Toolbar 组件属性

XML 属性	说明
app:navigationIcon	设置左侧图标
app:title	设置标题
app:titleTextColor	设置标题颜色
app:subtitle	设置子标题
app:subtitleTextColor	设置子标题颜色

而 DrawerLayout 则是 Google 推出的抽屉效果的组件，被 Google 包含在了 android-support-v4.jar 包中，可以让开发者更简单、方便地实现侧滑菜单这一功能。DrawerLayout 其实是一个布局组件，跟 LinearLayout 等组件是同一种东西，但带有滑动的功能。只要按照 DrawerLayout 的规定布局方式写完布局，就能产生侧滑的效果。

DrawerLayout 重要的布局特性如下。

◇ 在 DrawerLayout 中，主内容视图必须是第一个子视图，因为 XML 顺序意味着按 z 序（层叠顺序）排序，并且抽屉式导航栏必须位于内容顶部。z 序是指叠放次序（z-index），当对多个元素同时设定定位时，定位元素之间可能会发生重叠。z-index 默认值是 0，取值越大，叠放次序越高（即越靠上）。

◇ 主内容视图设置为匹配父视图的宽度和高度，因为在抽屉式导航栏处于隐藏状态时，它代表整个 UI。

◇ 抽屉式导航栏视图必须使用 android:layout_gravity 属性指定其为侧边栏布局。要支持"从右到左"（RTL）语言，请指定该值为"start"（而非"left"），这样当布局为 RTL 时，抽屉式导航栏会显示在右侧。

◇ 抽屉式导航栏视图以 dp 为单位指定其宽度，且高度与父视图匹配。抽屉式导航栏的宽度不应超过 320dp，从而使用户始终可以看到部分主内容。

NavigationView 表示应用程序的标准导航菜单，菜单内容可以由菜单（menu）资源文件填充。NavigationView 通常放在 DrawerLayout 中，可以实现侧滑效果的 UI。菜单整体分为两部分，上面的部分叫作 HeaderLayout，下面的菜单项都是 menu。NavigationView 组件常见的 XML 属性如表 13-2 所示。

表 13-2 NavigationView 组件属性

XML 属性	说明
android:layout_gravity	表示布局自身在父布局的哪个位置
app:headerLayout	表示引用一个头布局文件，用来展示 NavigationView 的头布局
app:menu	表示引用一个 menu 作为下面的菜单项

13.1.1 使用 ToolBar 组件

1)更改应用主题。

要使用 Toolbar 组件必须将原有的 Actionbar 隐藏,因此需要更改应用主题。主题在 styles.xml 文件中定义,如图 13-3 所示。

图 13-3 styles.xml 文件

更新后的 styles.xml 文件如下所示。

```
1. <resources>
2.     <style name="AppTheme" parent="Theme.AppCompat.Light.NoActionBar">
3.         <item name="android:textColorPrimary">@color/colorBlack</item>
4.         <item name="android:textSize">18sp</item>
5.     </style>
6. </resources>
```

2)在 builde.gradle 文件中添加包的依赖。

```
implementation 'com.android.support:appcompat-v7:28.0.0'
implementation 'com.android.support:design:28.0.0'
```

在 Android Studio 3.0 以上的运行环境导入,依赖包的关键字不再是"compile",而是"implementation"。

13.1.2 创建导航栏

1)新建 view_navigation_header 布局作为 NavigationView 的 HeaderLayout,效果如图 13-4 所示。

图 13-4 导航栏

view_navigation_header.xml 文件的代码如下。

```
1.  <?xml version="1.0" encoding="utf-8"?>
2.  <RelativeLayout
3.      xmlns:android="http://schemas.android.com/apk/res/android"
4.      android:layout_width="match_parent"
5.      android:layout_height="180dp"
6.      android:background="@color/colorLightBlue"
7.      android:orientation="vertical">
8.
9.      <TextView
10.         android:layout_width="wrap_content"
11.         android:layout_height="wrap_content"
12.         android:layout_marginLeft="15dp"
13.         android:layout_above="@+id/tv_accountName"
14.         android:text="UserName"
15.         android:textColor="@color/colorWhite"
16.         android:textSize="25sp"/>
17.     <TextView
18.         android:id="@+id/tv_accountName"
19.         android:layout_width="wrap_content"
20.         android:layout_height="wrap_content"
21.         android:layout_alignParentBottom="true"
22.         android:layout_margin="15dp"
23.         android:text="Account"
24.         android:textColor="@color/colorWhite"
```

```
25.         android:textSize="16sp"/>
26. </RelativeLayout>
```

2）在 menu 目录下新建 drawer_view.xml 文件作为 NavigationView 的目录，代码如下。

```
1. <?xml version="1.0" encoding="utf-8"?>
2. <menu xmlns:android="http://schemas.android.com/apk/res/android">
3.     <item android:title="@string/menu_item1">
4.         <menu>
5.             <item
6.                 android:id="@+id/setting_app"
7.                 android:icon="@mipmap/pic_setting"
8.                 android:title="@string/menu_item2"/>
9.             <item
10.                android:id="@+id/language_app"
11.                android:icon="@mipmap/pic_language"
12.                android:title="@string/menu_item3"/>
13.            <item
14.                android:id="@+id/about_app"
15.                android:icon="@mipmap/pic_about"
16.                android:title="@string/menu_item4"/>
17.        </menu>
18.    </item>
19.    <item android:title="@string/menu_item5">
20.        <menu>
21.            <item
22.                android:id="@+id/change_acc"
23.                android:icon="@mipmap/pic_cuser_w"
24.                android:title="@string/menu_item6"/>
25.            <item
26.                android:id="@+id/close_app"
27.                android:icon="@mipmap/pic_out"
28.                android:title="@string/menu_item7"/>
29.        </menu>
30.    </item>
31. </menu>
```

菜单元素放在<menu>标签下，分为两组 item，通过为 item 添加子菜单可以实现带有头部的分组效果，并且该组的最上面自动添加一条分隔线。item 中的 icon 及 title 用于设置图标以及标题。

3）更新 activity_main.xml 布局文件，代码如下。

```
1. <?xml version="1.0" encoding="utf-8"?>
2. <android.support.v4.widget.DrawerLayout
3.     xmlns:android="http://schemas.android.com/apk/res/android"
4.     xmlns:tools="http://schemas.android.com/tools"
5.     xmlns:app="http://schemas.android.com/apk/res-auto"
```

```
6.      android:id="@+id/main_drawer"
7.      android:layout_height="match_parent"
8.      android:layout_width="match_parent"
9.      android:fitsSystemWindows="true"
10.     tools:context=".MainActivity"
11.     >
12.     <LinearLayout
13.         xmlns:android="http://schemas.android.com/apk/res/android"
14.         xmlns:tools="http://schemas.android.com/tools"
15.         android:layout_width="match_parent"
16.         android:layout_height="match_parent"
17.         android:orientation="vertical"
18.         tools:context=".MainActivity"
19.         >
20.         <android.support.v7.widget.Toolbar
21.             android:id="@+id/toolbar"
22.             android:layout_width="match_parent"
23.             android:layout_height="?attr/actionBarSize"
24.             >
25.         </android.support.v7.widget.Toolbar>
26.         ...
27.     </LinearLayout>
28.     <android.support.design.widget.NavigationView
29.         android:id="@+id/main_nav"
30.         android:layout_width="match_parent"
31.         android:layout_height="match_parent"
32.         android:layout_gravity="start"
33.         app:menu="@menu/drawer_view"
34.         app:headerLayout="@layout/view_navigation_header"
35.         />
36. </android.support.v4.widget.DrawerLayout>
```

第 2~11 行 DrawerLayout 必须位于内容顶部，因此作为最外层的根布局。

第 9 行 fitsSystemWindows 属性可以让 view（即 DrawerLayout）根据系统窗口来调整自己的布局。

第 12~27 行在 DrawerLayout 内，添加一个包含屏幕主内容（当抽屉式导航栏处于隐藏状态时为主要布局）的布局。

第 20~25 行创建一个 Toolbar 组件用于替代 Actionbar，并通过第 23 行将高度设置为原有的高度。

第 28~35 行在 DrawerLayout 内创建另一个包含抽屉式导航栏内容的布局。

第 32 行指定侧滑栏从左向右滑出。

第 33 行引入菜单的布局 drawer_view。

第 34 行引入头布局文件 view_navigation_header。

4）在 MainActivity 中的 onCreate()方法添加代码实现抽屉导航的功能。

```
1.  navigationView=findViewById(R.id.main_nav);
2.  drawerLayout=findViewById(R.id.main_drawer);
3.  toolbar=findViewById(R.id.toolbar);
4.  setSupportActionBar(toolbar);
5.  //导航按钮
6.  ActionBarDrawerToggle mDrawerToggle=new ActionBarDrawerToggle(this,
7.  drawerLayout, toolbar, R.string.drawer_open, R.string.drawer_close);
8.  //同步状态
9.  mDrawerToggle.syncState();
10. drawerLayout.addDrawerListener(mDrawerToggle);
11. navigationView.setNavigationItemSelectedListener(
12. new NavigationView.OnNavigationItemSelectedListener() {
13.     @Override
14.     public boolean onNavigationItemSelected(@NonNull MenuItem item) {
15.         drawerLayout.closeDrawers();//单击侧滑菜单，关闭抽屉，隐藏菜单
16.         switch(item.getItemId()) {
17.         case R.id.setting_app:
18.             startActivity(new Intent(MainActivity.this,UserActivity.class));
19.             break;
20.         case R.id.language_app:
21.             startActivity(new Intent(MainActivity.this,LanguageActivity.class));
22.             break;
23.         case R.id.about_app:
24.             startActivity(new Intent(MainActivity.this,AboutActivity.class));
25.             break;
26.         case R.id.change_acc:
27.             break;
28.         case R.id.close_app:
29.             break;
30.         }
31.         return true;
32.     }
33. });
```

第1~3行初始化组件，方便接下来的操作。

第4行用 ToolBar 替代原有的 ActionBar。

第5~10行创建侧滑菜单状态监听器，监测 DrawLayout 的开关，并同步左侧图标状态。

第11~33行设置当导航栏被单击时的回调。通过这个回调关闭导航菜单以及其他操作。

第16~30行当单击菜单项时跳转到对应界面或进行相应的操作。

13.2 创建抽屉导航界面中各功能模块

抽屉导航界面中的功能模块包含用户信息展示、语言选择、软件信息、切换用户、退出程序等。

44 创建各功能模块

1）创建 view_title_frame.xml 文件，用于补充 ActionBar 的空缺，效果如图 13-5 所示。

图 13-5　view_title_frame 界面

view_title_frame.xml 文件的代码如下。

```
1.  <?xml version="1.0" encoding="utf-8"?>
2.  <RelativeLayout xmlns:android="http://schemas.android.com/apk/res/android"
3.      android:layout_width="wrap_content"
4.      android:layout_height="wrap_content"
5.      android:background="@color/colorLightBlue"
6.      android:fitsSystemWindows="true">
7.      <RelativeLayout
8.          android:layout_width="match_parent"
9.          android:layout_height="50dp">
10.         <TextView
11.             android:id="@+id/tv_back"
12.             android:layout_width="wrap_content"
13.             android:layout_height="wrap_content"
14.             android:layout_centerVertical="true"
15.             android:background="@drawable/pic_btn_back"
16.             android:gravity="center_vertical"
17.             android:text="@string/title_1"
18.             android:textColor="@color/colorWhite"
19.             android:textSize="18sp"/>
20.         <TextView
21.             android:id="@+id/tv_title"
```

```
22.            android:layout_width="wrap_content"
23.            android:layout_height="wrap_content"
24.            android:layout_centerInParent="true"
25.            android:text=""
26.            android:textColor="@color/colorWhite"
27.            android:textSize="18sp"/>
28.    </RelativeLayout>
29. </RelativeLayout>
```

2）创建用户信息展示界面布局文件 activity_user.xml，效果如图 13-6 所示。

图 13-6　用户信息展示界面

activity_user.xml 文件的代码如下。

```
1.  <?xml version="1.0" encoding="utf-8"?>
2.  <LinearLayout xmlns:android="http://schemas.android.com/apk/res/android"
3.      android:layout_width="match_parent"
4.      android:layout_height="match_parent"
5.      android:orientation="vertical">
6.      <include layout="@layout/view_title_frame"/>
7.      <android.support.v7.widget.CardView
8.          android:layout_width="match_parent"
9.          android:layout_height="wrap_content"
10.         android:layout_margin="5dp">
11.         <LinearLayout
12.             android:layout_width="match_parent"
```

```
13.            android:layout_height="wrap_content"
14.            android:orientation="vertical">
15.            <TextView
16.                android:layout_width="wrap_content"
17.                android:layout_height="wrap_content"
18.                android:layout_marginLeft="15dp"
19.                android:layout_marginTop="20dp"
20.                android:layout_marginBottom="10dp"
21.                android:gravity="center"
22.                android:text="@string/userlayout_text1"
23.                android:textColor="#959899"
24.                android:textSize="16sp"/>
25.            <View
26.                android:layout_width="match_parent"
27.                android:layout_height="1px"
28.                android:layout_marginLeft="15dp"
29.                android:layout_marginRight="15dp"
30.                android:background="#cecece"/>
31.            <RelativeLayout
32.                android:layout_width="match_parent"
33.                android:layout_height="48dp"
34.                android:paddingLeft="15dp"
35.                android:paddingRight="15dp">
36.                <TextView
37.                    android:id="@+id/title0"
38.                    android:layout_width="wrap_content"
39.                    android:layout_height="wrap_content"
40.                    android:layout_marginLeft="15dp"
41.                    android:layout_marginTop="20dp"
42.                    android:layout_marginBottom="10dp"
43.                    android:gravity="center"
44.                    android:text="@string/userlayout_text2"
45.                    android:textSize="16sp"/>
46.                <TextView
47.                    android:id="@+id/account_name"
48.                    android:layout_width="match_parent"
49.                    android:layout_height="match_parent"
50.                    android:layout_marginLeft="8dp"
51.                    android:layout_toRightOf="@+id/title0"
52.                    android:gravity="center_vertical"
53.                    android:paddingLeft="15dp"
54.                    android:textColor="#363636"
55.                    android:textSize="16sp"/>
56.                <View
57.                    android:layout_width="match_parent"
58.                    android:layout_height="1px"
59.                    android:layout_alignParentBottom="true"
```

```xml
60.                    android:layout_marginLeft="15dp"
61.                    android:layout_marginRight="15dp"
62.                    android:background="#cecece"/>
63.            </RelativeLayout>
64.            <RelativeLayout
65.                android:layout_width="match_parent"
66.                android:layout_height="48dp"
67.                android:paddingLeft="15dp"
68.                android:paddingRight="15dp">
69.                <TextView
70.                    android:id="@+id/title1"
71.                    android:layout_width="wrap_content"
72.                    android:layout_height="wrap_content"
73.                    android:layout_marginLeft="15dp"
74.                    android:layout_marginTop="20dp"
75.                    android:layout_marginBottom="10dp"
76.                    android:gravity="center"
77.                    android:text="@string/userlayout_text3"
78.                    android:textSize="16sp"/>
79.                <TextView
80.                    android:id="@+id/account_type"
81.                    android:layout_width="match_parent"
82.                    android:layout_height="match_parent"
83.                    android:layout_marginLeft="8dp"
84.                    android:layout_toRightOf="@+id/title1"
85.                    android:gravity="center_vertical"
86.                    android:paddingLeft="15dp"
87.                    android:textColor="#363636"
88.                    android:textSize="14sp"/>
89.                <View
90.                    android:layout_width="match_parent"
91.                    android:layout_height="1px"
92.                    android:layout_alignParentBottom="true"
93.                    android:layout_marginLeft="15dp"
94.                    android:layout_marginRight="15dp"
95.                    android:background="#cecece"/>
96.            </RelativeLayout>
97.            <RelativeLayout
98.                android:layout_width="match_parent"
99.                android:layout_height="48dp"
100.                android:paddingLeft="15dp"
101.                android:paddingRight="15dp">
102.                <TextView
103.                    android:id="@+id/title2"
104.                    android:layout_width="wrap_content"
105.                    android:layout_height="wrap_content"
106.                    android:layout_marginLeft="15dp"
```

```
107.                    android:layout_marginTop="20dp"
108.                    android:layout_marginBottom="10dp"
109.                    android:gravity="center"
110.                    android:text="@string/userlayout_text4"
111.                    android:textSize="16sp"/>
112.                <TextView
113.                    android:id="@+id/sex"
114.                    android:layout_width="match_parent"
115.                    android:layout_height="match_parent"
116.                    android:layout_marginLeft="8dp"
117.                    android:layout_toRightOf="@+id/title2"
118.                    android:gravity="center_vertical"
119.                    android:paddingLeft="15dp"
120.                    android:textColor="#363636"
121.                    android:textSize="14sp"/>
122.                <View
123.                    android:layout_width="match_parent"
124.                    android:layout_height="1px"
125.                    android:layout_alignParentBottom="true"
126.                    android:layout_marginLeft="15dp"
127.                    android:layout_marginRight="15dp"
128.                    android:background="#cecece"/>
129.            </RelativeLayout>
130.            <RelativeLayout
131.                android:layout_width="match_parent"
132.                android:layout_height="48dp"
133.                android:paddingLeft="15dp"
134.                android:paddingRight="15dp">
135.                <TextView
136.                    android:id="@+id/title3"
137.                    android:layout_width="wrap_content"
138.                    android:layout_height="wrap_content"
139.                    android:layout_marginLeft="15dp"
140.                    android:layout_marginTop="20dp"
141.                    android:layout_marginBottom="10dp"
142.                    android:gravity="center"
143.                    android:text="@string/userlayout_text5"
144.                    android:textSize="16sp"/>
145.                <TextView
146.                    android:id="@+id/phone_number"
147.                    android:layout_width="match_parent"
148.                    android:layout_height="match_parent"
149.                    android:layout_marginLeft="8dp"
150.                    android:layout_toRightOf="@+id/title3"
151.                    android:gravity="center_vertical"
152.                    android:paddingLeft="15dp"
153.                    android:textColor="#363636"
```

```
154.                    android:textSize="14sp"/>
155.                <View
156.                    android:layout_width="match_parent"
157.                    android:layout_height="1px"
158.                    android:layout_alignParentBottom="true"
159.                    android:layout_marginLeft="15dp"
160.                    android:layout_marginRight="15dp"
161.                    android:background="#cecece"/>
162.            </RelativeLayout>
163.            <RelativeLayout
164.                android:layout_width="match_parent"
165.                android:layout_height="48dp"
166.                android:paddingLeft="15dp"
167.                android:paddingRight="15dp">
168.                <TextView
169.                    android:id="@+id/title4"
170.                    android:layout_width="wrap_content"
171.                    android:layout_height="wrap_content"
172.                    android:layout_marginLeft="15dp"
173.                    android:layout_marginTop="20dp"
174.                    android:layout_marginBottom="10dp"
175.                    android:gravity="center"
176.                    android:text="@string/userlayout_text6"
177.                    android:textSize="16sp"/>
178.                <TextView
179.                    android:id="@+id/email"
180.                    android:layout_width="match_parent"
181.                    android:layout_height="match_parent"
182.                    android:layout_marginLeft="8dp"
183.                    android:layout_toRightOf="@+id/title4"
184.                    android:gravity="center_vertical"
185.                    android:paddingLeft="15dp"
186.                    android:textColor="#363636"
187.                    android:textSize="16sp"/>
188.                <View
189.                    android:layout_width="match_parent"
190.                    android:layout_height="1px"
191.                    android:layout_alignParentBottom="true"
192.                    android:layout_marginLeft="15dp"
193.                    android:layout_marginRight="15dp"
194.                    android:background="#cecece"/>
195.            </RelativeLayout>
196.            <RelativeLayout
197.                android:layout_width="match_parent"
198.                android:layout_height="48dp"
199.                android:paddingLeft="15dp"
200.                android:paddingRight="15dp">
```

```
201.            <TextView
202.                android:id="@+id/title5"
203.                android:layout_width="wrap_content"
204.                android:layout_height="wrap_content"
205.                android:layout_marginLeft="15dp"
206.                android:layout_marginTop="20dp"
207.                android:layout_marginBottom="10dp"
208.                android:gravity="center"
209.                android:text="@string/userlayout_text7"
210.                android:textSize="16sp"/>
211.            <TextView
212.                android:layout_width="match_parent"
213.                android:layout_height="match_parent"
214.                android:layout_marginLeft="8dp"
215.                android:layout_toRightOf="@+id/title5"
216.                android:gravity="center_vertical"
217.                android:paddingLeft="15dp"
218.                android:textColor="#363636"
219.                android:textSize="16sp"/>
220.            <View
221.                android:layout_width="match_parent"
222.                android:layout_height="1px"
223.                android:layout_alignParentBottom="true"
224.                android:layout_marginLeft="15dp"
225.                android:layout_marginRight="15dp"
226.                android:background="#cecece"/>
227.        </RelativeLayout>
228.        <RelativeLayout
229.            android:layout_width="match_parent"
230.            android:layout_height="48dp"
231.            android:paddingLeft="15dp"
232.            android:paddingRight="15dp">
233.            <TextView
234.                android:id="@+id/title6"
235.                android:layout_width="wrap_content"
236.                android:layout_height="wrap_content"
237.                android:layout_marginLeft="15dp"
238.                android:layout_marginTop="20dp"
239.                android:layout_marginBottom="10dp"
240.                android:gravity="center"
241.                android:text="@string/userlayout_text8"
242.                android:textSize="16sp"/>
243.            <TextView
244.                android:layout_width="match_parent"
245.                android:layout_height="match_parent"
246.                android:layout_marginLeft="8dp"
247.                android:layout_toRightOf="@+id/title6"
```

```xml
248.                    android:gravity="center_vertical"
249.                    android:paddingLeft="15dp"
250.                    android:textColor="#363636"
251.                    android:textSize="16sp"/>
252.                <View
253.                    android:layout_width="match_parent"
254.                    android:layout_height="1px"
255.                    android:layout_alignParentBottom="true"
256.                    android:layout_marginLeft="15dp"
257.                    android:layout_marginRight="15dp"
258.                    android:background="#cecece"/>
259.            </RelativeLayout>
260.            <RelativeLayout
261.                android:layout_width="match_parent"
262.                android:layout_height="48dp"
263.                android:paddingLeft="15dp"
264.                android:paddingRight="15dp">
265.                <TextView
266.                    android:id="@+id/title7"
267.                    android:layout_width="wrap_content"
268.                    android:layout_height="wrap_content"
269.                    android:layout_marginLeft="15dp"
270.                    android:layout_marginTop="20dp"
271.                    android:layout_marginBottom="10dp"
272.                    android:gravity="center"
273.                    android:text="@string/userlayout_text9"
274.                    android:textSize="16sp"/>
275.                <TextView
276.                    android:layout_width="match_parent"
277.                    android:layout_height="match_parent"
278.                    android:layout_marginLeft="8dp"
279.                    android:layout_toRightOf="@+id/title7"
280.                    android:gravity="center_vertical"
281.                    android:paddingLeft="15dp"
282.                    android:textColor="#363636"
283.                    android:textSize="16sp"/>
284.                <View
285.                    android:layout_width="match_parent"
286.                    android:layout_height="1px"
287.                    android:layout_alignParentBottom="true"
288.                    android:layout_marginLeft="15dp"
289.                    android:layout_marginRight="15dp"
290.                    android:background="#cecece"/>
291.            </RelativeLayout>
292.        </LinearLayout>
```

```
293.        </android.support.v7.widget.CardView>
294.    </LinearLayout>
```

外层使用 LinearLayout 嵌套了 CardView。CardView 继承自 FrameLayout，是一种扁平化视图，配合属性能够实现漂亮的卡片效果。CardView 组件常见的属性如表 13-3 所示。

表 13-3　CardView 组件属性

XML 属性	说明
app:cardBackgroundColor	设置背景颜色
app:cardCornerRadius	设置圆角大小
app:cardElevation	设置 z 轴阴影
app:contentPadding	设置内容的内边距，即 CardView 子布局与 CardView 边界的距离

第 6 行在 include 中导入了 view_title_frame 布局，可以实现界面的返回，提高了共同布局的复用性。

第 7~293 行设置 CardView 内部的布局。

第 15~30 行添加 CardView 的标题，并设置分隔线。

第 31~63 行设置 CardView 一行的布局内容。接下来每行的布局基本与此相同。

3）创建 UserActivity 活动，代码如下。

```
1.  package cn.edu.jsit.smartfactory;
2.  import android.annotation.SuppressLint;
3.  import android.content.SharedPreferences;
4.  import android.os.Bundle;
5.  import android.os.Handler;
6.  import android.os.Message;
7.  import android.support.annotation.Nullable;
8.  import android.support.v4.app.ActivityCompat;
9.  import android.view.View;
10. import android.widget.TextView;
11. import org.json.JSONArray;
12. import org.json.JSONException;
13. import org.json.JSONObject;
14. import cn.edu.jsit.smartfactory.tools.WebServiceHelper;
15. public class UserActivity extends AppCompatActivity {
16.     TextView ac_tv,sex_tv,phone_tv,e_tv;
17.     String ac,sex,phone,email,result;
18.     SharedPreferences spPreferences;
19.     @SuppressLint("HandlerLeak")
20.     Handler handler=new Handler() {
21.         @Override
22.         public void handleMessage(Message msg) {
23.             super.handleMessage(msg);
24.             try {
25.                 //根据服务器返回信息解析JSON数据
```

```java
26.                JSONArray array=new JSONArray(result);
27.                JSONObject object=(JSONObject) array.get(0);
28.                sex=object.getString("sys_Sex");
29.                phone=object.getString("sys_Phone");
30.                email=object.getString("sys_Email");
31.                ac_tv.setText(ac);
32.                sex_tv.setText(sex);
33.                phone_tv.setText(phone);
34.                e_tv.setText(email);
35.            } catch(JSONException e) {
36.                e.printStackTrace();
37.            }
38.        }
39.    };
40.    @Override
41.    protected void onCreate(@Nullable Bundle savedInstanceState) {
42.        super.onCreate(savedInstanceState);
43.        setContentView(R.layout.activity_user);
44.        ((TextView) findViewById(R.id.account_type)).setText("普通账号");
45.        ac_tv=(TextView) findViewById(R.id.account_name);
46.        sex_tv=(TextView) findViewById(R.id.sex);
47.        phone_tv=(TextView) findViewById(R.id.phone_number);
48.        e_tv=(TextView) findViewById(R.id.email);
49.        spPreferences=getSharedPreferences("loginSet", MODE_PRIVATE);
50.        ac=spPreferences.getString("user", "");
51.        WebServiceHelper.GetById(new WebServiceHelper.Callback() {
52.            @Override
53.            public void call(String s) {
54.                result=s;
55.                handler.sendEmptyMessage(0);
56.            }
57.        },ac);
58.        setListener();
59.    }
60.    public void setListener() {
61.        ((TextView) findViewById(R.id.tv_back)).setOnClickListener(new View.OnClickListener() {
62.            @Override
63.            public void onClick(View v) {
64.                ActivityCompat.finishAfterTransition(UserActivity.this);
65.            }
66.        });
67.    }
68. }
```

第 20~39 行实现 handler 的内部匿名类，处理从 WebService 中获取的数据。

第 26~34 行解析 JSON 格式的数据,并将数据显示在界面中。将从服务器端获取的字符串以数组形式存入 JSONArray,再使用 JSONObject 解析出单个的对象,最后使用 getString()方法通过对象名称获取对应的值,并将值设置在界面上。

JSON 是一种轻量级的数据交换格式,采用文本格式存储和表示数据,具有良好的可读性和便于快速编写的特性。JSON 的语法规则如下。

✧ 数据定义在名称/值对中。
✧ 数据由逗号分隔。
✧ 花括号保存对象。
✧ 方括号保存数组。

① JSON 数据的书写格式是名称/值对。名称/值对包括字段名称(在双引号中),后面写一个冒号,然后是值,例如:

```
{"Name" : "John"}
```

等价于 Java 语句:Name="John"。

② JSON 值可以是数字(整型或浮点型)、字符串(在双引号中)、逻辑值(true 或 false)、数组(在方括号中)、对象(在花括号中)和 null。

③ JSON 数字可以是整型或浮点型,例如:

```
{ "age":30 }
```

④ JSON 对象在花括号中书写,对象可以包含多个名称/值对,之间用逗号分隔,例如:

```
{ "firstName":"John" , "lastName":"Doe" }
```

一组{}就是一个 JSON 对象。

JSON 数组在方括号中书写,数组可包含多个对象,例如:

```
1.  {
2.    "employees":
3.    [
4.      { "firstName":"John" , "lastName":"Doe" },
5.      { "firstName":"Anna" , "lastName":"Smith" },
6.      { "firstName":"Peter" , "lastName":"Jones" }
7.    ]
8.  }
```

在上面的例子中,对象 employees 是包含三个对象的数组。每个对象代表一条关于某个员工(firstName、lastName)的信息。

第 40~59 行初始化界面,通过 WebService 获取服务器端的用户信息。用 GetById()方法传入账号,即可获取该账号注册时的所有信息。使用 handler 将数据发送到主界面中去更新。

第 49、50 行读取登录时 SharePreferences 保存的账号。

第 51 行中的 GetById()方法将在本书任务 14 中创建。

第 60~67 添加单击事件,回到主界面中。

4)创建语言选择界面布局文件 activity_language.xml,效果如图 13-7 所示。

图 13-7 语言选择界面

activity_language.xml 文件的代码如下。

```
1.  <?xml version="1.0" encoding="utf-8"?>
2.  <LinearLayout xmlns:android="http://schemas.android.com/apk/res/android"
3.      android:layout_width="match_parent"
4.      android:layout_height="match_parent"
5.      android:orientation="vertical">
6.      <include layout="@layout/view_title_frame"/>
7.      <ListView
8.          android:id="@+id/lv_language"
9.          android:layout_width="match_parent"
10.         android:layout_height="match_parent"
11.         android:layout_marginTop="25dp"
12.         android:dividerHeight="1px"/>
13. </LinearLayout>
```

5）创建 LanguageActivity 活动，代码如下。

```
1.  package cn.edu.jsit.smartfactory;
2.  import android.content.Intent;
3.  import android.content.SharedPreferences;
4.  import android.os.Bundle;
5.  import android.view.KeyEvent;
6.  import android.view.View;
7.  import android.widget.AdapterView;
```

```
8.  import android.widget.ListView;
9.  import android.widget.TextView;
10. import cn.edu.jsit.smartfactory.adapter.LanguageAdapter;
11. import cn.edu.jsit.smartfactory.tools.SmartFactoryApplication;
12. public class LanguageActivity extends AppCompatActivity {
13.     private TextView tv_back, tv_title;
14.     private ListView lv_language;
15.     private SharedPreferences spPreferences;
16.     private LanguageAdapter languageAdapter;
17.     private String newLanguage;
18.     @Override
19.     protected void onCreate(Bundle savedInstanceState) {
20.         super.onCreate(savedInstanceState);
21.         setContentView(R.layout.activity_language);
22.         initView();
23.     }
24.     @Override
25.     public boolean onKeyDown(int keyCode, KeyEvent event) {
26.         switch(keyCode) {
27.         case KeyEvent.KEYCODE_BACK:
28.             closeActivity();
29.             return true;
30.         default:
31.             break;
32.         }
33.         return super.onKeyDown(keyCode, event);
34.     }
35.     private void closeActivity() {
36.         if(!newLanguage.equals(SmartFactoryApplication.language)) {
37.             SmartFactoryApplication.language=newLanguage;
38.             Intent intent=new Intent(this, MainActivity.class);
39.             intent.setFlags(Intent.FLAG_ACTIVITY_NEW_TASK |
40.             Intent.FLAG_ACTIVITY_CLEAR_TASK);
41.             startActivity(intent);
42.         }
43.         finish();
44.     }
45.     private void initView() {
46.         tv_back=(TextView) findViewById(R.id.tv_back);
47.         //返回后退
48.         tv_back.setOnClickListener(new View.OnClickListener() {
49.             @Override
50.             public void onClick(View v) {
51.                 closeActivity();
52.             }
53.         });
```

```
54.         tv_title=(TextView) findViewById(R.id.tv_title);
55.         tv_title.setText(R.string.language_title);
56.         lv_language=(ListView) findViewById(R.id.lv_language);
57.     }
58. }
```

第 24～44 行实现当单击返回按钮时，回到主界面。

第 36～42 行中，intent.setFlags()方法为当前意图 intent 设置了若干了 flags，其意义如下。

① FLAG_ACTIVITY_NEW_TASK。当 Intent 对象包含这个标记时，系统会寻找或创建一个新的 task 来放置目标 Activity，寻找时依据目标 Activity 的 taskAffinity 属性进行匹配。如果找到一个 task 的 taskAffinity 与之相同，就将目标 Activity 压入此 task 中；如果查找无果，则创建一个新的 task，并将该 task 的 taskAffinity 设置为目标 Activity 的 taskActivity，将目标 Activity 放置于此 task。注意，如果同一个应用中 Activity 的 taskAffinity 都使用默认值或都设置相同值时，应用内的 Activity 之间的跳转使用这个标记是没有意义的，因为当前应用 task 就是目标 Activity 最好的宿主。

② FLAG_ACTIVITY_CLEAR_TASK。如果在调用 Context.startActivity 时传递这个标记，将会导致任何用来放置该 Activity 的已经存在的 task 里面的已经存在的 Activity 先清空，然后该 Activity 再在该 task 中启动。也就是说，这个新启动的 Activity 变为了这个空 task 的根 Activity，所有老的 Activity 都结束。该标记必须和 FLAG_ACTIVITY_NEW_TASK 一起使用。

第 45～57 行初始化组件，添加返回单击事件。

6）创建关于软件界面布局文件 activity_about.xml，效果如图 13-8 所示。

图 13-8　关于软件界面

activity_about.xml 文件的代码如下。

```
1.  <?xml version="1.0" encoding="utf-8"?>
2.  <LinearLayout xmlns:android="http://schemas.android.com/apk/res/android"
3.      android:layout_width="match_parent"
4.      android:layout_height="match_parent"
5.      android:orientation="vertical">
6.      <include layout="@layout/view_title_frame"/>
7.      <ScrollView
8.          android:layout_width="match_parent"
9.          android:layout_height="match_parent"
10.         android:background="#f5f5f5"
11.         android:fillViewport="true">
12.         <RelativeLayout
```

```xml
13.         android:id="@+id/frame"
14.         android:layout_width="match_parent"
15.         android:layout_height="match_parent">
16.         <ImageView
17.             android:id="@+id/img"
18.             android:layout_width="70dp"
19.             android:layout_height="70dp"
20.             android:layout_centerHorizontal="true"
21.             android:layout_margin="15dp"
22.             android:src="@mipmap/icon_launcher"/>
23.         <TextView
24.             android:id="@+id/app_name"
25.             android:layout_width="wrap_content"
26.             android:layout_height="wrap_content"
27.             android:layout_below="@+id/img"
28.             android:layout_centerHorizontal="true"
29.             android:text="@string/app_name"
30.             android:textSize="20sp"/>
31.         <TextView
32.             android:id="@+id/version_id"
33.             android:layout_width="wrap_content"
34.             android:layout_height="wrap_content"
35.             android:layout_below="@+id/app_name"
36.             android:layout_centerHorizontal="true"
37.             android:layout_marginTop="5dp"
38.             android:text="V 1.0"
39.             android:textSize="14sp"/>
40.         <LinearLayout
41.             android:id="@+id/buttonPanel"
42.             android:layout_width="match_parent"
43.             android:layout_height="match_parent"
44.             android:layout_below="@+id/version_id"
45.             android:layout_above="@+id/bottom_btn"
46.             android:orientation="vertical">
47.         </LinearLayout>
48.         <LinearLayout
49.             android:id="@+id/bottom_btn"
50.             android:layout_width="wrap_content"
51.             android:layout_height="wrap_content"
52.             android:layout_above="@+id/bottom_text"
53.             android:layout_centerHorizontal="true"
54.             android:layout_marginBottom="5dp"
55.             android:visibility="invisible"
56.             android:orientation="horizontal">
57.         </LinearLayout>
```

```xml
58.            <TextView
59.                android:id="@+id/bottom_text"
60.                android:layout_width="wrap_content"
61.                android:layout_height="wrap_content"
62.                android:layout_alignParentBottom="true"
63.                android:layout_centerHorizontal="true"
64.                android:layout_marginBottom="10dp"
65.                android:text="@string/copyRight_text"
66.                android:textSize="10sp" />
67.        </RelativeLayout>
68.    </ScrollView>
69. </LinearLayout>
```

7）创建 AboutActivity 活动，代码如下。

```java
1.  package cn.edu.jsit.smartfactory;
2.  import android.os.Bundle;
3.  import android.support.v4.app.ActivityCompat;
4.  import android.view.View;
5.  import android.widget.TextView;
6.  public class AboutActivity extends AppCompatActivity {
7.      private TextView version_info;
8.      @Override
9.      protected void onCreate(Bundle savedInstanceState) {
10.         super.onCreate(savedInstanceState);
11.         setContentView(R.layout.activity_about);
12.         initView();
13.     }
14.     void initView() {
15.         version_info=(TextView) findViewById(R.id.version_id);
16.         version_info.setText(R.string.version);
17.         ((TextView) findViewById(R.id.tv_title)).setText("关于");
18.         //返回后退
19.         ((TextView) findViewById(R.id.tv_back)).
20.         setOnClickListener(new View.OnClickListener() {
21.             @Override
22.             public void onClick(View view) {
23.                 ActivityCompat.finishAfterTransition(AboutActivity.this);
24.             }
25.         });
26.     }
27. }
```

第 6 行创建 AboutActivity 类并继承 AppCompatActivity。

第 20～25 行创建单击事件，回到主界面中。

第 23 行实现页面过渡动画，从当前界面退出到主界面。

8）补充 MainActivity 中的代码，完成切换账号以及退出程序的功能。

```java
1.  public void logOut() {
2.      dialog.setTitle(R.string.account_out);
3.      dialog.setMessage(R.string.toast_3);
4.      dialog.setPositiveButton(R.string.btn_ok, new DialogInterface.OnClickListener() {
5.          @Override
6.          public void onClick(DialogInterface arg0, int arg1) {
7.              Intent intent=new Intent();
8.              intent.setClass(MainActivity.this, LoginActivity.class);
9.              startActivity(intent);
10.         }
11.     });
12.     dialog.setNegativeButton(R.string.btn_cancel, null);
13.     dialog.show();
14. }
15. public void exitDialog() {
16.     dialog.setTitle(R.string.btn_out);
17.     dialog.setMessage(R.string.toast_4);
18.     dialog.setPositiveButton(R.string.btn_ok,new DialogInterface.OnClickListener() {
19.         @Override
20.         public void onClick(DialogInterface arg0, int arg1) {
21.             finish();
22.             System.exit(0);
23.         }
24.     });
25.     dialog.setNegativeButton(R.string.btn_cancel, null);
26.     dialog.show();
27. }
```

logOut()方法为切换账号功能，当单击弹出对话框中的"确定"按钮时，程序会进入登录界面。此时可以重新填写账号信息。setTitle()方法用于设置对话框的标题，setMessage()方法用于设置对话框的提示信息。setPositiveButton()以及 setNegativeButton()方法分别用于设置对话框下方两侧的按钮及监听单击事件，并在其回调方法中可以执行操作。Show()方法用于显示对话框。

exitDialog()方法为退出程序功能，当单击弹出对话框中的"确定"按钮时，判断当前是否登录了云平台，是则退出云平台后再退出程序，否则直接退出。

任务 14 创建登录和注册功能

任务概述

登录界面包含了 APP 的图标、名称、账号和密码输入框、登录和注册按钮，如图 14-1 所示。用户如已注册账号，则可在输入框中输入账号、密码，然后单击"登录"按钮，程序会向服务器端验证用户信息，通过则可进入主界面。如用户未注册账号，可以通过单击下方的"注册账号"按钮进入注册界面，如图 14-2 所示。在注册界面中需要用户填写 5 个信息，其中账号和密码为必填项，填写完成后单击"注册完成"按钮返回主界面（单击上方的"返回"也可以回到主界面），程序会将用户填写的信息保存至服务器端的数据库中，在个人设置界面可以查看注册的信息。

图 14-1　登录界面　　　　　图 14-2　注册界面

知识目标

● 掌握 RadioButton 组件。
● 掌握 RadioGroup 组件。

技能目标

● 能熟练使用自定义样式实现系统组件的美化。

14.1 创建并部署 WebService

45 创建和部署报警信息管理 WebService

本任务中的 WebService 服务使用 C#语言编写，开发环境为 Visual Studio 2017（如果对 C#不了解，可以忽略该内容，不会影响本书的阅读）。

服务器端的 WebService 代码如下。

```csharp
1.  using System;
2.  using System.Collections.Generic;
3.  using System.Linq;
4.  using System.Web;
5.  using System.Web.Services;
6.  using System.Data;
7.  using System.Data.SqlClient;
8.  using Newtonsoft.Json;
9.  ///<summary>
10. ///appservice 的摘要说明
11. ///</summary>
12. [WebService(Namespace="http://tempuri.org/")]
13. [WebServiceBinding(ConformsTo=WsiProfiles.BasicProfile1_1)]
14. //若要允许使用 ASP.NET AJAX 从脚本中调用此 Web 服务，请取消注释下一行
15. //[System.Web.Script.Services.ScriptService]
16. public class appservice:System.Web.Services.WebService
17. {
18.     public appservice()
19.     {
20.         //如果使用设计的组件，请取消注释下一行
21.         //InitializeComponent();
22.     }
23.     public string strcon="data source=.;initial catalog=AppService;user id=sa;password=123456";
24.     [WebMethod]
25.     public string register(string name,string password,string sex,string phone,string email)
26.     {
27.         try
28.         {
29.             using(SqlConnection con=new SqlConnection(strcon))
30.             {
31.                 con.Open();
32.                 SqlCommand cmd=new SqlCommand("insert into sys_UserInfo values(@sys_Name,@sys_Password,@sys_Sex,@sys_Phone,@sys_Email)", con);
33.                 cmd.Parameters.Add("sys_Name", SqlDbType.NVarChar).Value=name;
34.                 cmd.Parameters.Add("sys_Password", SqlDbType.NVarChar).Value=password;
```

```
35.                cmd.Parameters.Add("sys_Sex", SqlDbType.NVarChar).Value=sex;
36.                cmd.Parameters.Add("sys_Phone",SqlDbType.NVarChar).Value=phone;
37.                cmd.Parameters.Add("sys_Email",SqlDbType.NVarChar).Value=email;
38.                cmd.ExecuteNonQuery();
39.            }
40.            return "注册成功！";
41.        }
42.        catch(Exception ex)
43.        {
44.            return "注册失败！\r\n 错误信息："+ex.ToString();
45.        }
46.    }
47.    [WebMethod]
48.    public string Login(string name,string password)
49.    {
50.        try
51.        {
52.            using(SqlConnection con=new SqlConnection(strcon))
53.            {
54.                con.Open();
55.                SqlCommand cmd=new SqlCommand("select * from sys_UserInfo where sys_Name=@name and sys_Password=@password", con);
56.                cmd.Parameters.Add("name", SqlDbType.NVarChar).Value=name;
57.                cmd.Parameters.Add("password",SqlDbType.NVarChar).Value=password;
58.                SqlDataReader re=cmd.ExecuteReader();
59.                if(re.HasRows)
60.                {
61.                    return "true";
62.                }
63.            }
64.            return "false";
65.        }
66.        catch(Exception ex)
67.        {
68.            return ex.ToString();
69.        }
70.    }
71.    [WebMethod]
72.    public string getAll(string sys_Name)
73.    {
74.        List<Info> info=new List<Info>();
75.        string result="";
76.        SqlConnection conn=new SqlConnection(strcon);
```

```
77.         conn.Open();
78.         SqlCommand cmd=new SqlCommand("select * from sys_UserInfo where sys_Name=@id", conn);
79.         cmd.Parameters.Add("id", SqlDbType.NVarChar).Value=sys_Name;
80.         SqlDataReader dr=cmd.ExecuteReader();
81.         if(dr.Read())
82.         {
83.             Info a=new Info();
84.             a.sys_Sex=dr[3].ToString();
85.             a.sys_Phone=dr[4].ToString();
86.             a.sys_Email=dr[5].ToString();
87.             info.Add(a);
88.         }
89.         result=JsonConvert.SerializeObject(info);
90.         dr.Close();
91.         conn.Close();
92.         return result;
93.     }
94.     public class Info
95.     {
96.         public string sys_Sex { get; set; }
97.         public string sys_Phone { get; set; }
98.         public string sys_Email { get; set; }
99.     }
100. }
```

第 24~46 行定义了 register()方法。

第 47~70 行定义了 login()方法。

第 71~93 行定义了 getAll()方法。

14.2 在 WebServiceHelper 类中添加登录和注册功能

46 添加登录和注册功能

在 WebServiceHelper 类中添加登录和注册功能的代码如下。

```
1.  public static void GetLogin(final Callback callback,final String user,final String psw){
2.      new Thread(){
3.          public void run() {
4.              SoapSerializationEnvelope envelope=new SoapSerializationEnvelope(SoapEnvelope.VER10);
5.              SoapObject rpc=new SoapObject("http://tempuri.org/", "Login");
6.              rpc.addProperty("name", user);
7.              rpc.addProperty("password", psw);
8.              envelope.bodyOut=rpc;
9.              envelope.dotNet=true;
10.             envelope.setOutputSoapObject(rpc);
```

```
11.             HttpTransportSE transportSE=new HttpTransportSE("http://192.
168.0.2:9001/appservice.asmx?wsdl");
12.             try {
13.                 transportSE.call("http://tempuri.org/Login", envelope);
14.                 SoapObject object=new SoapObject();
15.                 object=(SoapObject) envelope.bodyIn;
16.                 String result=object.getProperty(0).toString();
17.                 Log.i("result", result);
18.                 callback.call(result);
19.             } catch(IOException e) {
20.                 //TODO Auto-generated catch block
21.                 e.printStackTrace();
22.             } catch(XmlPullParserException e) {
23.                 //TODO Auto-generated catch block
24.                 e.printStackTrace();
25.             }
26.         };
27.     }.start();
28. }
29. public static void SetReg(final Callback callback,final String user,final String psw,final String sex,final String phone,final String email){
30.     new Thread(){
31.         public void run() {
32.             SoapSerializationEnvelope envelope=new SoapSerializationEnvelope(SoapEnvelope.VER10);
33.             SoapObject rpc=new SoapObject("http://tempuri.org/", "register");
34.             rpc.addProperty("name", user);
35.             rpc.addProperty("password", psw);
36.             rpc.addProperty("sex", sex);
37.             rpc.addProperty("phone", phone);
38.             rpc.addProperty("email", email);
39.             envelope.bodyOut=rpc;
40.             envelope.dotNet=true;
41.             envelope.setOutputSoapObject(rpc);
42.             HttpTransportSE transportSE=new HttpTransportSE("http://192.
168.0.2:9001/appservice.asmx?wsdl");
43.             try {
44.                 transportSE.call("http://tempuri.org/register", envelope);
45.                 SoapObject object=new SoapObject();
46.                 object=(SoapObject) envelope.bodyIn;
47.                 String result=object.getProperty(0).toString();
48.                 Log.i("result", result);
49.                 callback.call(result);
50.             } catch(IOException e) {
51.                 //TODO Auto-generated catch block
52.                 e.printStackTrace();
53.             } catch (XmlPullParserException e) {
```

```
54.                //TODO Auto-generated catch block
55.                e.printStackTrace();
56.            }
57.        };
58.    }.start();
59. }
60. public static void GetById(final Callback callback,final String id){
61.     new Thread(){
62.         public void run() {
63.             SoapSerializationEnvelope envelope=new SoapSerializationEnvelope(SoapEnvelope.VER10);
64.             SoapObject rpc=new SoapObject("http://tempuri.org/", "getAll");
65.             rpc.addProperty("sys_Name", id);
66.             envelope.bodyOut=rpc;
67.             envelope.dotNet=true;
68.             envelope.setOutputSoapObject(rpc);
69.             HttpTransportSE transportSE=new HttpTransportSE("http://192.168.0.2:9001/appservice.asmx?wsdl");
70.             try {
71.                 transportSE.call("http://tempuri.org/getAll", envelope);
72.                 SoapObject object=new SoapObject();
73.                 object=(SoapObject) envelope.bodyIn;
74.                 String result=object.getProperty(0).toString();
75.                 Log.i("result", result);
76.                 callback.call(result);
77.             } catch(IOException e) {
78.                 //TODO Auto-generated catch block
79.                 e.printStackTrace();
80.             } catch(XmlPullParserException e) {
81.                 //TODO Auto-generated catch block
82.                 e.printStackTrace();
83.             }
84.         };
85.     }.start();
86. }
```

在 WebServiceHelper 类中，新添加了三个方法 GetLogin()、SetReg()和 GetById()，分别用于登录账号、注册账号和获取账号注册信息。

WebService 的基本用法与之前的任务相同，首先新建 SoapSerializationEnvelope 对象生成调用 WebService 方法的 SOAP 信息，并且指定 SOAP 版本。再新建 SoapObject 对象指定 WebService 的命名空间和调用方法。addProperty()方法设置需要调用 WebService 接口的参数。bodyOut()与 setOutputSoapObject()方法作用相同，提交请求信息，传出 SOAP 消息体。由于 WebService 是.NET 开发的，所以这里 dotNet 要设置为 true 来兼容.NET 服务器端。构建 HttpTransportSE 传输对象，并指定 WDSL 文档中的 URL，该对象用于调用 WebService 操作。call()方法中的参数为：命名空间、方法名称和 envelope 对象用于调用 WebService。接下来调用完成，访问 SoapSerializationEnvelope 对象的 bodyIn 属性，该属性返回一个 SoapObject 对象，该对象代表 WebService 的返回信息。最后将返回信息读取并转化为字符串形式保存，传递到回调方法 Callback()中。

14.3 创建登录和注册界面

14.3.1 创建登录界面

1）创建登录界面的布局文件 activity_login.xml，效果如图 14-3 所示。

图 14-3 登录界面

activity_login.xml 文件的代码如下。

```
1.  <?xml version="1.0" encoding="utf-8"?>
2.  <RelativeLayout
3.      xmlns:android="http://schemas.android.com/apk/res/android"
4.      android:layout_width="match_parent"
5.      android:layout_height="match_parent">
6.      <ImageView
7.          android:id="@+id/img_icon"
8.          android:layout_width="100dp"
9.          android:layout_height="100dp"
10.         android:layout_centerHorizontal="true"
11.         android:layout_marginTop="80dp"
12.         android:src="@mipmap/icon_launcher"
```

```xml
13.            />
14.        <TextView
15.            android:id="@+id/tv_name"
16.            android:layout_width="wrap_content"
17.            android:layout_height="wrap_content"
18.            android:textSize="24sp"
19.            android:textColor="@color/colorPrimaryDark"
20.            android:layout_centerHorizontal="true"
21.            android:layout_below="@id/img_icon"
22.            android:layout_marginTop="10dp"
23.            android:text="@string/app_name"
24.            />
25.        <EditText
26.            android:id="@+id/account_id"
27.            android:layout_width="match_parent"
28.            android:layout_height="38dp"
29.            android:layout_marginLeft="38dp"
30.            android:layout_marginRight="38dp"
31.            android:layout_marginTop="35dp"
32.            android:layout_below="@id/tv_name"
33.            android:padding="5dp"
34.            android:background="@drawable/bg_edittextview"
35.            android:drawableLeft="@mipmap/pic_edit_user"
36.            android:drawablePadding="8dp"
37.            android:hint="@string/account_id_account"
38.            android:imeOptions="actionNext"
39.            android:inputType="textEmailAddress"
40.            android:maxLength="11"
41.            android:singleLine="true"
42.            android:textColor="#000000"
43.            android:textSize="16sp"/>
44.        <requestFocus/>
45.        <EditText
46.            android:id="@+id/account_psw"
47.            android:layout_width="match_parent"
48.            android:layout_height="38dp"
49.            android:layout_marginLeft="38dp"
50.            android:layout_marginRight="38dp"
51.            android:layout_marginTop="10dp"
52.            android:layout_below="@id/account_id"
53.            android:drawablePadding="8dp"
54.            android:padding="5dp"
55.            android:background="@drawable/bg_edittextview"
56.            android:drawableLeft="@mipmap/pic_edit_pwd"
57.            android:hint="@string/account_id_password"
58.            android:imeOptions="actionDone"
59.            android:maxLength="16"
```

```
60.        android:inputType="textPassword"
61.        android:singleLine="true"
62.        android:textColor="#000000"
63.        android:textSize="16sp"/>
64.    <requestFocus/>
65.    <Button
66.        android:id="@+id/btn_loginUp"
67.        android:layout_width="match_parent"
68.        android:layout_height="38dp"
69.        android:layout_marginLeft="38dp"
70.        android:layout_marginRight="38dp"
71.        android:layout_marginTop="35dp"
72.        android:layout_below="@id/account_psw"
73.        android:background="@drawable/bg_login_blue"
74.        android:text="@string/account_loginUp"
75.        android:textSize="18sp"
76.        android:textColor="@color/colorWhite"/>
77.    <TextView
78.        android:id="@+id/tv_registered"
79.        android:layout_width="wrap_content"
80.        android:layout_height="wrap_content"
81.        android:layout_marginTop="10dp"
82.        android:text="@string/account_register"
83.        android:layout_below="@id/btn_loginUp"
84.        android:layout_centerHorizontal="true"
85.        android:textColor="#000000"/>
86. </RelativeLayout>
```

EditText 组件在前面的任务中使用过，这里用于用户输入账号和密码，然后获取用户输入的内容，提交给服务器进行判断。EditText 组件的一些其他属性如表 14-1 所示。

表 14-1 EditText 组件其他属性

XML 属性	说明
android:textColorHint	为空时显示的文本的颜色
android:inputType	限制输入类型 number：整数类型 numberDecimal：小数点类型 date：日期类型 text：文本类型（默认值） phone：拨号键盘 textPassword：密码 textVisiblePassword：可见密码 textUrl：网址
android:maxLength	限制显示的文本长度，超出部分不显示
android:singleLine	true 为单行显示，false 为可以多行
android:imeOptions	输入法 Enter 键图标的设置

2）Android 系统原生的 EditText 组件外形并不美观，因此我们需要自己给 EditText 定义一个圆角蓝边的样式，如图 14-4 所示。

图 14-4 圆角蓝边样式

在 drawbale 目录下新建 bg_edt_bluestroke.xml 文件，设置 EditText 未选中时的圆角样式，代码如下。

```
1.  <?xml version="1.0" encoding="utf-8"?>
2.  <shape xmlns:android="http://schemas.android.com/apk/res/android">
3.      <stroke
4.          android:width="1dp"
5.          android:color="@color/text_blue"/>
6.      <corners
7.          android:radius="9dp"
8.          android:topLeftRadius="12dp"
9.          android:topRightRadius="12dp"
10.         android:bottomLeftRadius="12dp"
11.         android:bottomRightRadius="12dp"/><!-- 设置圆角半径 -->
12.     <solid android:color="@color/colorWhite"/>
13. </shape>
```

3）在 drawbale 目录下新建 bg_edt_bluestroke_press.xml 文件，设置 EditText 被选中时的圆角样式，代码如下。

```
1.  <?xml version="1.0" encoding="utf-8"?>
2.  <shape xmlns:android="http://schemas.android.com/apk/res/android">
3.      <stroke
4.          android:width="1dp"
5.          android:color="@color/grey21"/>
6.      <corners
7.          android:radius="9dp"
8.          android:topLeftRadius="12dp"
9.          android:topRightRadius="12dp"
10.         android:bottomLeftRadius="12dp"
11.         android:bottomRightRadius="12dp"/><!-- 设置圆角半径 -->
12.     <solid android:color="@color/background_hui"/>
13. </shape>
```

4）在 drawbale 目录下新建 bg_edittextview.xml 文件，代码如下。

```xml
1. <?xml version="1.0" encoding="utf-8"?>
2. <selector xmlns:android="http://schemas.android.com/apk/res/android">
3.     <item
4.         android:state_pressed="false"
5.         android:drawable="@drawable/bg_edt_bluestroke"/>
6.     <item
7.         android:state_pressed="true"
8.         android:drawable="@drawable/bg_edt_bluestroke_press"/>
9.     <item
10.         android:drawable="@drawable/bg_edt_bluestroke"/>
11. </selector>
```

实际应用中，很多地方如 Button、Tab、ListItem 等都是在不同状态下有不同的展示形状。例如，一个按钮的背景，默认时是一个形状，按下时是一个形状，不可操作时又是另一个形状。因此前面分别定义了 EditText 选中与未选中状态时的两种样式。有时候，不同状态下改变的不只是背景、图片等，文字颜色也会相应改变。而要处理这些不同状态下展示不同形状的问题，就要用 selector 来实现。

在给 EditText 定义完成样式以后需要将样式设置给 EditText。

```
android:background="@drawable/bg_edittextview"
```

14.3.2 创建注册界面

创建注册界面布局文件 activity_register.xml，效果如图 14-5 所示。

图 14-5 注册界面

activity_register.xml 文件的代码如下。

```xml
1.  <?xml version="1.0" encoding="utf-8"?>
2.  <RelativeLayout xmlns:android="http://schemas.android.com/apk/res/android"
3.      android:layout_width="match_parent"
4.      android:layout_height="match_parent">
5.      <include layout="@layout/view_title_frame"/>
6.      <TextView
7.          android:id="@+id/name"
8.          android:layout_width="wrap_content"
9.          android:layout_height="wrap_content"
10.         android:layout_marginTop="120dp"
11.         android:layout_centerHorizontal="true"
12.         android:layout_marginBottom="85dp"
13.         android:text="@string/account_register"
14.         android:textSize="30sp"/>
15.     <LinearLayout
16.         android:layout_width="match_parent"
17.         android:layout_height="wrap_content"
18.         android:focusable="true"
19.         android:focusableInTouchMode="true"
20.         android:orientation="vertical"
21.         android:layout_alignParentBottom="true"
22.         android:layout_marginTop="-55dp"
23.         android:layout_below="@+id/name"
24.         >
25.         <EditText
26.             android:id="@+id/registered_id"
27.             android:layout_width="match_parent"
28.             android:layout_height="38dp"
29.             android:layout_marginLeft="38dp"
30.             android:layout_marginRight="38dp"
31.             android:padding="5dp"
32.             android:background="@drawable/bg_edittextview"
33.             android:hint="@string/register_account"
34.             android:gravity="center"
35.             android:imeOptions="actionNext"
36.             android:maxLength="11"
37.             android:singleLine="true"
38.             android:textColor="#000000"
39.             android:textSize="16sp"/>
40.         <requestFocus/>
41.         <EditText
42.             android:id="@+id/registered_psw"
43.             android:layout_width="match_parent"
44.             android:layout_height="38dp"
45.             android:layout_marginLeft="38dp"
```

```
46.            android:layout_marginRight="38dp"
47.            android:layout_marginTop="15dp"
48.            android:padding="5dp"
49.            android:background="@drawable/bg_edittextview"
50.            android:hint="@string/register_password"
51.            android:gravity="center"
52.            android:imeOptions="actionNext"
53.            android:maxLength="16"
54.            android:singleLine="true"
55.            android:textColor="#000000"
56.            android:textSize="16sp"/>
57.        <requestFocus/>
58.        <EditText
59.            android:id="@+id/registered_phone"
60.            android:layout_width="match_parent"
61.            android:layout_height="38dp"
62.            android:layout_marginLeft="38dp"
63.            android:layout_marginRight="38dp"
64.            android:layout_marginTop="15dp"
65.            android:padding="5dp"
66.            android:background="@drawable/bg_edittextview"
67.            android:hint="@string/register_phone"
68.            android:gravity="center"
69.            android:imeOptions="actionNext"
70.            android:inputType="phone"
71.            android:maxLength="16"
72.            android:singleLine="true"
73.            android:textColor="#000000"
74.            android:textSize="16sp"/>
75.        <requestFocus/>
76.        <EditText
77.            android:id="@+id/registered_email"
78.            android:layout_width="match_parent"
79.            android:layout_height="38dp"
80.            android:layout_marginLeft="38dp"
81.            android:layout_marginRight="38dp"
82.            android:layout_marginTop="15dp"
83.            android:padding="5dp"
84.            android:background="@drawable/bg_edittextview"
85.            android:hint="@string/register_email"
86.            android:gravity="center"
87.            android:inputType="textEmailAddress"
88.            android:imeOptions="actionDone"
89.            android:maxLength="16"
90.            android:singleLine="true"
91.            android:textColor="#000000"
```

```
92.            android:textSize="16sp"/>
93.        <requestFocus/>
94.        <RadioGroup
95.            android:id="@+id/rg_sex"
96.            android:layout_width="match_parent"
97.            android:layout_height="38dp"
98.            android:layout_marginLeft="38dp"
99.            android:layout_marginRight="38dp"
100.            android:layout_marginTop="15dp"
101.            android:orientation="horizontal"
102.            android:gravity="center"
103.            >
104.            <RadioButton
105.                android:id="@+id/rg_male"
106.                android:layout_width="wrap_content"
107.                android:layout_height="wrap_content"
108.                android:text="@string/sex_male"
109.                />
110.            <RadioButton
111.                android:id="@+id/rg_female"
112.                android:layout_width="wrap_content"
113.                android:layout_height="wrap_content"
114.                android:text="@string/sex_female"
115.                android:layout_marginLeft="30dp"
116.                />
117.        </RadioGroup>
118.        <Button
119.            android:id="@+id/btn_registered"
120.            android:layout_width="match_parent"
121.            android:layout_height="38dp"
122.            android:layout_marginLeft="38dp"
123.            android:layout_marginRight="38dp"
124.            android:layout_marginTop="25dp"
125.            android:background="@drawable/bg_login_blue"
126.            android:text="@string/register_ok"
127.            android:textSize="18sp"
128.            android:textColor="@color/colorWhite"/>
129.    </LinearLayout>
130. </RelativeLayout>
```

RadioButton 即单选按钮，是一种基础的 UI 组件。RadioGroup 提供了 RadioButton 单选按钮的容器，RadioButton 通常放于 RadioGroup 容器中使用。

要实现 RadioButton 的功能，通常需要由 RadioButton 和 RadioGroup 配合使用。其中 RadioGroup 是单选组合框，可以容纳多个 RadioButton 的。在没有 RadioGroup 的情况下，RadioButton 可以全部被选中；当多个 RadioButton 被 RadioGroup 包含的情况下，RadioButton 只可以选择一个，从而达到单选的目的。使用 setOnCheckChangeLinstener()方法

来对单选按钮进行监听。RadioButton 组件的属性如表 14-2 所示。

表 14-2 RadioButton 组件属性

XML 属性	说明
android:drawable	设置图片，可以选择图片位置
android:checked	表示组件是否选中
android:button	隐藏按钮前面的圆圈

RadioGroup 的监听单击事件代码如下。

```
1. RadioGroup.setOnCheckedChangeListener(new RadioGroup.OnCheckedChangeListener() {
2.     @Override
3.     public void onCheckedChanged(RadioGroup group, int checkedId) {
4.         switch(checkedId){
5.             case R.id.xx:
6.                 showToast(R.string.xx);
7.                 break;
8.             case R.id.xx:
9.                 showToast(R.string.xx);
10.                 break;
11.         }
12.     }
13. });
```

setOnCheckedChangeListener 实现内部匿名类并复写 onCheckedChanged() 方法。当 RadioButton 被单击时，程序回调 onCheckedChanged。checkedId 即是 RadioGroup 组内的 RadioButton 的 id，根据 id 对相应的 RadioButton 进行操作。

14.4 创建 LoginActivity 活动实现登录功能

LoginActivity.java 文件的代码如下。

48 实现登录功能

```
1. package cn.edu.jsit.smartfactory;
2. import android.annotation.SuppressLint;
3. import android.content.Intent;
4. import android.content.SharedPreferences;
5. import android.os.Bundle;
6. import android.os.Handler;
7. import android.os.Message;
8. import android.support.annotation.Nullable;
9. import android.support.v7.app.AppCompatActivity;
10. import android.text.TextUtils;
11. import android.view.View;
12. import android.widget.Button;
13. import android.widget.EditText;
14. import android.widget.TextView;
15. import android.widget.Toast;
```

```java
16.     import cn.edu.jsit.smartfactory.tools.WebServiceHelper;
17.     public class LoginActivity extends AppCompatActivity {
18.         private EditText ed_id,ed_psw;
19.         private TextView tv_register;
20.         private Button btn_login;
21.         private SharedPreferences spPreferences;
22.         String result;
23.         @SuppressLint("HandlerLeak")
24.         Handler handler=new Handler(){
25.             @Override
26.             public void handleMessage(Message msg) {
27.                 if(msg.what==1){
28.                     SharedPreferences.Editor editor=spPreferences.edit();
29.                     editor.putString("pwd", ed_psw.getText().toString().trim());
30.                     editor.putString("user", ed_id.getText().toString().trim());
31.                     editor.commit();
32.                     Intent intent=new Intent();
33.                     intent.setClass(LoginActivity.this, MainActivity.class);
34.                     startActivity(intent);
35.                     finish();
36.                 }else{
37.                     Toast.makeText(LoginActivity.this, R.string.account_toast_text2, Toast.LENGTH_SHORT).show();
38.                 }
39.             }
40.         };
41.         @Override
42.         protected void onCreate(@Nullable Bundle savedInstanceState) {
43.             super.onCreate(savedInstanceState);
44.             setContentView(R.layout.activity_login);
45.             ed_id=findViewById(R.id.account_id);
46.             ed_psw=findViewById(R.id.account_psw);
47.             tv_register=findViewById(R.id.tv_registered);
48.             btn_login=findViewById(R.id.btn_loginUp);
49.             spPreferences=getSharedPreferences("loginSet", MODE_PRIVATE);
50.             ed_id.setText(spPreferences.getString("user", ""));
51.             ed_psw.setText(spPreferences.getString("pwd", ""));
52.             tv_register.setOnClickListener(new View.OnClickListener() {
53.                 @Override
54.                 public void onClick(View view) {
55.                     Intent intent=new Intent(LoginActivity.this,RegisterActivity.class);
56.                     startActivity(intent);
57.                 }
58.             });
59.             btn_login.setOnClickListener(new View.OnClickListener() {
60.                 @Override
```

```
61.         public void onClick(View view) {
62.             loginUp(ed_id.getEditableText().toString(), ed_psw.getEditableText().toString());
63.         }
64.     });
65. }
66. private void loginUp(String userName, String passWord) {
67.     if(TextUtils.isEmpty(userName)||TextUtils.isEmpty(passWord)) {
68.         Toast.makeText(LoginActivity.this, R.string.account_toast_text1, Toast.LENGTH_SHORT).show();
69.         return;
70.     }
71.     WebServiceHelper.GetLogin(new WebServiceHelper.Callback(){
72.         @Override
73.         public void call(String s) {
74.             result=s;
75.             if(result.equals("true")){
76.                 handler.sendEmptyMessage(1);
77.             }else{
78.                 handler.sendEmptyMessage(0);
79.             }
80.         }
81.     },userName,passWord);
82. }
83. }
```

第 24~40 行新建 handler 对象处理网络请求后返回的数据。

第 28~31 行登录成功则默认保存登录时的账号信息，方便下次直接登录。

第 32~35 行新建 intent 意图跳转到主界面并结束当前的登录界面。

第 43~58 行初始化界面及组件。

第 49~51 行读取保存的账号信息并显示在界面上。

第 52~58 行添加跳转到注册界面的单击事件。

第 59~64 行添加登录按钮的单击事件，将输入框中的账号信息作为参数传入 loginUp()方法中。

第 66~82 行新建了 loginUp()方法用于账号的登录。

第 67、68 行判断当前的账号、密码输入框是否为空，只要其中有一个为空则提示警告信息。

第 71~81 行调用 WebServiceHelper 的 GetLogin()方法传入参数访问服务器端，并接收服务器端返回的信息，判断是否登录成功。

14.5 创建 RegisterActivity 活动实现注册功能

RegisterActivity.java 文件的代码如下。

49 实现注册功能

```java
1.  package cn.edu.jsit.smartfactory;
2.  import android.os.Bundle;
3.  import android.os.Handler;
4.  import android.os.Message;
5.  import android.support.annotation.Nullable;
6.  import android.support.v4.app.ActivityCompat;
7.  import android.support.v7.app.AppCompatActivity;
8.  import android.view.View;
9.  import android.widget.Button;
10. import android.widget.EditText;
11. import android.widget.RadioGroup;
12. import android.widget.TextView;
13. import android.widget.Toast;
14. import cn.edu.jsit.smartfactory.tools.WebServiceHelper;
15. public class RegisterActivity extends AppCompatActivity {
16.     private Button registered_button;
17.     private EditText id_edit,psw_edit,phone_edit,email_edit;
18.     private String id,psw,phone,sex,email,result;
19.     private RadioGroup radioGroup;
20.     Handler handler=new Handler(){
21.         @Override
22.         public void handleMessage(Message msg) {
23.             Toast.makeText(RegisterActivity.this,
24.                     result, Toast.LENGTH_SHORT).show();
25.             finish();
26.         }
27.     };
28.     @Override
29.     protected void onCreate(@Nullable Bundle savedInstanceState) {
30.         super.onCreate(savedInstanceState);
31.         setContentView(R.layout.activity_register);
32.         registered_button=(Button) findViewById(R.id.btn_registered);
33.         id_edit=(EditText) findViewById(R.id.registered_id);
34.         psw_edit=(EditText) findViewById(R.id.registered_psw);
35.         phone_edit=(EditText) findViewById(R.id.registered_phone);
36.         email_edit=(EditText) findViewById(R.id.registered_email);
37.         radioGroup=findViewById(R.id.rg_sex);
38.         ((TextView) findViewById(R.id.tv_back)).setOnClickListener(new View.OnClickListener() {
39.             @Override
40.             public void onClick(View v) {
41.                 ActivityCompat.finishAfterTransition(RegisterActivity.this);
42.             }
43.         });
44.         radioGroup.setOnCheckedChangeListener(new RadioGroup.OnCheckedChangeListener() {
```

```
45.        @Override
46.        public void onCheckedChanged(RadioGroup radioGroup, int i) {
47.            switch(radioGroup.getCheckedRadioButtonId()){
48.                case R.id.rg_male:
49.                    sex="男";
50.                    break;
51.                case R.id.rg_female:
52.                    sex="女";
53.                    break;
54.            }
55.        }
56.    });
57.    registered_button.setOnClickListener(new View.OnClickListener() {
58.        @Override
59.        public void onClick(View view) {
60.            id=id_edit.getText().toString().trim();
61.            psw=psw_edit.getText().toString().trim();
62.            phone=phone_edit.getText().toString().trim();
63.            email=email_edit.getText().toString().trim();
64.            if(id.equals("") && psw.equals("")) {
65.                Toast.makeText(RegisterActivity.this,
66.                    R.string.account_toast_text1, Toast.LENGTH_SHORT).show();
67.            } else {
68.                //向服务器提交注册信息
69.                WebServiceHelper.SetReg(new WebServiceHelper.Callback() {
70.                    @Override
71.                    public void call(String s) {
72.                        result=s;
73.                        handler.sendEmptyMessage(0);
74.                    }
75.                }, id, psw, sex, phone, email);
76.            }
77.        }
78.    });
79.  }
80. }
```

第20～27行新建handler对象处理服务器返回的结果。

第28～79行初始化界面及组件，实现单击事件。

第38～43行添加返回主界面的单击事件，finishAfterTransition为界面切换的过渡方式。

第44～56行添加RadioGroup的单击事件，获取用户选择的性别信息。

50 完善注销和退出功能

第57～78行添加提交注册信息的单击事件，单击时从各组件中读取用户填写的信息，用户名及密码不能为空，否则弹出提示信息。调用WebService的GetReg()方法将获取的用户信息作为参数传入，最后处理服务器端返回的结果。

任务 15　实现多语言切换

任务概述

本任务在任务 13 的基础上，完善 Android 应用多语言切换的功能。进入语言选择界面，显示当前选择的语言选项，单击其他选项即可切换至相应的语言，如图 15-1 所示。回到主界面，可以看到界面语言已经更改成功，如图 15-2 所示。

图 15-1　语言选择界面

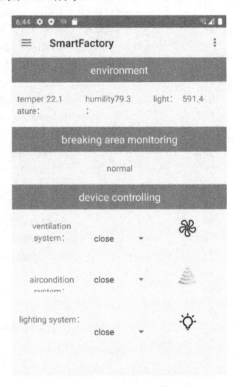

图 15-2　切换语言后的主界面

知识目标

● 掌握 Configuration 类。

技能目标

● 能实现 Android 应用多语言的切换。

51　更新语言选择界面

15.1　更新语言选择界面

在任务 13 中的 activity_language.xml 布局文件中，我们使用了 ListView 组件，这里需要在 ListView 中填充内容，完成使用 ListView 选择语言的功能。

15.1.1 创建语言选择界面

创建语言选择布局文件 items_lv_language.xml，效果如图 15-3 所示。

图 15-3 语言选择界面

items_lv_language.xml 文件的代码如下。

```
1.  <?xml version="1.0" encoding="utf-8"?>
2.  <RelativeLayout xmlns:android="http://schemas.android.com/apk/res/android"
3.      android:layout_width="match_parent"
4.      android:layout_height="wrap_content">
5.      <RelativeLayout
6.          android:layout_width="match_parent"
7.          android:layout_height="45dp"
8.          >
9.          <TextView
10.             android:id="@+id/language_data"
11.             android:layout_height="wrap_content"
12.             android:layout_width="wrap_content"
13.             android:layout_centerVertical="true"
14.             android:layout_marginLeft="15dp"
15.             android:text=""/>
16.         <ImageView
17.             android:id="@+id/img_isCheck"
18.             android:layout_width="30dp"
19.             android:layout_height="30dp"
20.             android:layout_alignParentRight="true"
21.             android:layout_centerVertical="true"
```

```
22.            android:layout_marginRight="15dp"
23.            android:src="@mipmap/pic_ok"/>
24.     </RelativeLayout>
25. </RelativeLayout>
```

15.1.2 创建 ListView 适配器

1）创建 LanguageAdapter.java 文件作为 ListView 的适配器。

```
1.  package cn.edu.jsit.smartfactory.adapter;
2.  import android.content.Context;
3.  import android.view.LayoutInflater;
4.  import android.view.View;
5.  import android.view.ViewGroup;
6.  import android.widget.BaseAdapter;
7.  import android.widget.ImageView;
8.  import android.widget.TextView;
9.  import cn.edu.jsit.smartfactory.R;
10. public class LanguageAdapter extends BaseAdapter {
11.     private Context context;
12.     private String[] data;
13.     private ViewHolder mHolder;
14.     private int isCheck=-1;
15.     public LanguageAdapter(Context context, String[] data) {
16.         this.context=context;
17.         this.data=data;
18.     }
19.     public void setCheck(int pos){
20.         isCheck=pos;
21.         notifyDataSetChanged();
22.     }
23.     @Override
24.     public int getCount() {
25.         return data.length;
26.     }
27.     @Override
28.     public Object getItem(int position) {
29.         return data[position];
30.     }
31.     @Override
32.     public long getItemId(int position) {
33.         return position;
34.     }
35.     @Override
36.     public View getView(int position, View convertView, ViewGroup parent) {
37.         if(convertView==null) {
38.             convertView=LayoutInflater.from(context)
```

```
39.            .inflate(R.layout.items_lv_language, null);
40.            mHolder=new ViewHolder();
41.            mHolder.language_data=(TextView) convertView
42.                .findViewById(R.id.language_data);
43.            mHolder.img=(ImageView) convertView.findViewById(R.id.img_is Check);
44.            //将 ViewHolder 存储在 View 中
45.            convertView.setTag(mHolder);
46.        } else {
47.            //重新获取 ViewHolder
48.            mHolder=(ViewHolder) convertView.getTag();
49.        }
50.        mHolder.language_data.setText(data[position]);
51.        if(isCheck==position) {
52.            mHolder.img.setImageDrawable(context.getResources()
53.                .getDrawable(R.mipmap.pic_ok));
54.        } else {
55.            mHolder.img.setImageDrawable(null);
56.        }
57.        return convertView;
58.    }
59.    private class ViewHolder {
60.        TextView language_data;
61.        ImageView img;
62.    }
63. }
```

新建 LanguageAdapter 类继承自适配器的基础类 BaseAdapter，并实现了抽象方法 getCount()、getItem()、getItemId()及 getView()。getCount()方法返回适配器中数据集的数据个数。getItem()方法获取数据集与索引对应的数据项。getItemId()方法获取指定行对应的 id。getView()方法获取每一行 item 的显示内容。

第15~18 行创建构造方法，用于接收传递的参数。

第35~58 行利用 ListView 的缓存机制，使用 ViewHolder 类实现显示数据视图的缓存，避免多次调用 findViewById 来寻找组件，以达到优化程序的目的。

第37~49 行判断 convertView 是否为空，为空则创建并初始化 ViewHolder 并设置到 tag；不为空则通过 tag 取出 ViewHolder。

第50~56 行给 ViewHolder 的组件设置数据。判断当前 item 是否已被单击，未被单击则将 ImageView 设置为空，已单击则显示勾选图标。

第59~62 行中的 ViewHolder 类用于缓存组件，两个属性分别对应 item 布局的组件。

2）更新 LanguageActivity 活动中的代码，为 ListView 设置数据适配器。

```
1. String[] data=new String[]{getString(R.string.language_default), getString(R.string.language_zh), getString(R.string.language_en_rUS)};
2. languageAdapter=new LanguageAdapter(this, data);
3. lv_language.setAdapter(languageAdapter);
```

15.2 简体中文和英文语言适配

52 简繁体中文语言适配

因为需要多种语言，所以要在项目中声明所需要的语言资源文件。

在工程的 res 目录下，新建 values-en 文件夹（后缀表示语言，比如-en 表示英文，-fr 表示法语，-es 表示西班牙语，等等），在这个文件夹目录下，新建字符串资源 XML 文件 strings.xml。

1）在 res 目录下新建 values-en-rUS 与 values-zh-rCN 文件夹，如图 15-4 所示。

...\app\src\main\res**values**
...\app\src\main\res**values-en-rUS**
...\app\src\main\res**values-zh-rCN**

图 15-4　语言文件夹

2）在两个文件夹中分别创建 strings.xml 文件并输入相应的文本，如图 15-5 所示。

图 15-5　strings.xml 文件

文件创建完成后，就可以简单地输入一些中英文对应的字符串资源了。例如，"跳过"这个字符资源，首先在 values 目录下的 strings.xml 文件中存入"跳过"的中文字符串资源，并赋予 name="skip_welcome"，如下所示。

```
1. <?xml version="1.0" encoding="utf-8"?>
2. <resources>
3.     <string name="skip_welcome">跳过</string>
4. </resources>
```

在 values-en-rUS 目录下的 strings.xml 文件中存入对应 name 的英文字符串资源，如下所示。

```
1. <?xml version="1.0" encoding="utf-8"?>
2. <resources>
3.     <string name="skip_welcome">skip</string>
4. </resources>
```

到这里就已经实现了简单的中英文适配。当需要使用"跳过"字符资源时，用 R.string.skip_welcome 即可调用该字符串资源。当切换系统语言环境的时候，程序会自动调用 values-en-rUS 目录下的 strings.xml 文件中 id 为 skip_welcome 的资源。

15.3 实现 Android 应用内切换语言

53 实现 Android 应用内切换语言

1)添加相关资源文件,并引用。在不同的 values 文件夹下(例如 values、values-en-rUS、values-zh-rCN 文件夹)添加不同语言的 strings.xml 文件,之前已经完成此步骤。

2)更新应用语言。更新 Android 中 Configuration 类中的 Locale 属性,可以实现对应用语言的变更,如下所示。

```
1. Configuration configuration=getResources().getConfiguration();
2. configuration.locale=Locale.ENGLISH;
3. getResources().updateConfiguration(configuration, null);
```

第 1 行获取 Configuration 对象。
第 2 行设置语言。
第 3 行更新 Configuration 对象。

但对于不熟悉其他语言的人,如果不小心切换了语言,并且回到了主界面,再想设置回中文可能就比较麻烦。所以 APP 最好能够在切换语言后立即刷新并且停留在当前的界面。因此我们创建 BaseActivity 类,将所有 Activity 文件继承 BaseActivity。具体实现方法如下。

```
1.  package cn.edu.jsit.smartfactory;
2.  import android.content.res.Configuration;
3.  import android.content.res.Resources;
4.  import android.os.Bundle;
5.  import android.support.v7.app.AppCompatActivity;
6.  import android.util.DisplayMetrics;
7.  import java.util.Locale;
8.  public class BaseActivity extends AppCompatActivity {
9.      @Override
10.     protected void onCreate(Bundle savedInstanceState) {
11.         super.onCreate(savedInstanceState);
12.     }
13.     public void changeLanguage(String language) {
14.         Locale myLocale;
15.         switch(language) {
16.             case "zh":
17.                 //简体中文
18.                 myLocale=Locale.SIMPLIFIED_CHINESE;
19.                 break;
20.             // 英文
21.             case "en_rUS":
22.                 myLocale=Locale.English;
23.                 break;
24.             default:
25.                 myLocale=Locale.getDefault();
26.                 break;
```

```
27.        }
28.        //根据Configuration的locale属性来加载语言的string资源
29.        Resources res=getResources();
30.        DisplayMetrics dm=res.getDisplayMetrics();
31.        Configuration conf=res.getConfiguration();
32.        conf.locale=myLocale;
33.        res.updateConfiguration(conf, dm);
34.    }
35. }
```

BaseActivity类继承自AppCompatActivity并实现onCreate()方法。

第13～34行创建changeLanguage()方法，根据传入的参数给myLocale赋值，再加载相应语言的string资源。

第29～33行更新Configuration中的Locale属性，在Configuration中指定语言类型。Configuration包含了设备的所有配置信息，这些信息会影响应用获取的资源。例如string资源，就是根据Configuration的locale属性来判断该取哪种语言的string资源，默认是在values文件夹下的。此处使用了Locale的预设值Locale.SIMPLIFIED_CHINESE，表示简体中文。跟随系统设置是Locale.getDefault()。

3）重启应用。选择某种语言后会将string资源文件替换为选择的语言，应用并不会自动刷新当前已经打开的Activity，所以为了刷新整个应用，需要重新启动一下应用。更新的LanguageActivity活动代码如下。

```
1.  package cn.edu.jsit.smartfactory;
2.  import android.content.Intent;
3.  import android.content.SharedPreferences;
4.  import android.os.Bundle;
5.  import android.view.KeyEvent;
6.  import android.view.View;
7.  import android.widget.AdapterView;
8.  import android.widget.ListView;
9.  import android.widget.TextView;
10. import cn.edu.jsit.smartfactory.adapter.LanguageAdapter;
11. import cn.edu.jsit.smartfactory.tools.SmartFactoryApplication;
12. public class LanguageActivity extends BaseActivity {
13.     private TextView tv_back, tv_title;
14.     private ListView lv_language;
15.     private SharedPreferences spPreferences;
16.     private LanguageAdapter languageAdapter;
17.     private String newLanguage;
18.     @Override
19.     protected void onCreate(Bundle savedInstanceState) {
20.         super.onCreate(savedInstanceState);
21.         setContentView(R.layout.activity_language);
22.         initView();
23.         initAdapter();
```

```
24.        }
25.        @Override
26.        public boolean onKeyDown(int keyCode, KeyEvent event) {
27.            switch(keyCode) {
28.                case KeyEvent.KEYCODE_BACK:
29.                    closeActivity();
30.                    return true;
31.
32.                default:
33.                    break;
34.            }
35.            return super.onKeyDown(keyCode, event);
36.        }
37.        private void closeActivity() {
38.            if(!newLanguage.equals(SmartFactoryApplication.language)) {
39.                SmartFactoryApplication.language=newLanguage;
40.                Intent intent=new Intent(this, MainActivity.class);
41.                intent.setFlags(Intent.FLAG_ACTIVITY_NEW_TASK | Intent.FLAG_ACTIVITY_CLEAR_TASK);
42.                startActivity(intent);
43.            }
44.            finish();
45.        }
46.        private void initView() {
47.            tv_back=(TextView) findViewById(R.id.tv_back);
48.            //返回后退
49.            tv_back.setOnClickListener(new View.OnClickListener() {
50.                @Override
51.                public void onClick(View v) {
52.                    closeActivity();
53.                }
54.            });
55.            tv_title=(TextView) findViewById(R.id.tv_title);
56.            tv_title.setText(R.string.language_title);
57.            lv_language=(ListView) findViewById(R.id.lv_language);
58.        }
59.        private void initAdapter() {
60.            String[]data=new String[]{getString(R.string.language_default), getString(R.string.language_zh), getString(R.string.language_en_rUS)};
61.            languageAdapter=new LanguageAdapter(this, data);
62.            lv_language.setAdapter(languageAdapter);
63.            lv_language.setOnItemClickListener(new AdapterView.OnItemClickListener() {
64.                @Override
65.                public void onItemClick(AdapterView<?>parent, View view, int position, long id) {
```

```
66.                    languageAdapter.setCheck(position);
67.                    switch(position) {
68.                        case 0:
69.                            setLanguage("default");
70.                            changeLanguage("default");
71.                            break;
72.                        case 1:
73.                            setLanguage("zh");
74.                            changeLanguage("zh");
75.                            break;
76.                        case 2:
77.                            setLanguage("en_rUS");
78.                            changeLanguage("en_rUS");
79.                            break;
80.                    }
81.                    LanguageActivity.this.finish();
82.                    LanguageActivity.this.
83.                        startActivity(new Intent(LanguageActivity.this, LanguageActivity.class));
84.                }
85.            });
86.            getLanguage();
87.        }
88.        public void getLanguage() {
89.            spPreferences=getSharedPreferences("loginSet", MODE_PRIVATE);
90.            newLanguage=spPreferences.getString("language", "default");
91.            switch(newLanguage) {
92.                case "zh":
93.                    languageAdapter.setCheck(1);
94.                    break;
95.                case "en_rUS":
96.                    languageAdapter.setCheck(2);
97.                    break;
98.                default:
99.                    languageAdapter.setCheck(0);
100.                    break;
101.            }
102.        }
103.        public void setLanguage(String language) {
104.            SharedPreferences.Editor editor=spPreferences.edit();
105.            editor.putString("language", language);
106.            editor.commit();
107.        }
108.    }
```

第37～43行让程序切换完语言后返回主界面。

第 59~87 行创建了 initAdapter()方法，为 ListView 设置数据适配器。setOnItemClickListener()方法监听 ListView 的单击事件。当某个 item 被单击时，首先适配器刷新界面，保存当前的选项，然后根据 item 内容更改语言，重新启动界面。

第 88~102 行创建 getLanguage()方法获取之前选择的语言，刷新界面使其显示为上次的选项。

第 103~107 行保存选择的语言选项。

4）在 SlpashActivity 中添加代码，保证程序在退出后再次启动能够显示退出时的语言。

```
1.  spPreferences=getSharedPreferences("loginSet", MODE_PRIVATE);
2.  SmartFactoryApplication.language=spPreferences.getString("language","default");
3.  changeLanguage(SmartFactoryApplication.language);
```

从 SharedPreferences 中读取保存的语言选项，并调用 changeLanguage()方法切换语言的 string 资源文件。

参 考 文 献

[1] 施密特. Java 完全参考手册[M]. 王德才,吴明飞,唐业军,译. 8版. 北京:清华大学出版社,2012.

[2] 郭霖. 第一行代码 Android[M]. 2版. 北京:人民邮电出版社,2016.

[3] Dawn Griffiths,David Griffiths. Head First Android 开发[M]. 乔莹,刘海洋,等译. 2版. 北京:中国电力出版社,2018.

[4] 陈继欣,等. 传感网应用开发职业技能等级标准[S]. 北京:北京新大陆时代教育科技有限公司,2019.